Leaving Earth:

Why one-way to Mars makes sense

by Andrew Rader

Ph.D., Aerospace Engineering (MIT)

Published on CreateSpace

Copyright © 2014 by Andrew Rader
132 pages
Cambridge, MA
United States of America
ISBN-13: 978-1495358975
ISBN-10: 1495358976
BISAC: Technology & Engineering / Aeronautics & Astronautics

To my friend Iain, who taught me that we can't just sit around and wait for the future – we have to create it.

If you want to build a ship, don't drum up people to collect wood, and don't assign them tasks and work, but rather teach them to long for the endless immensity of the sea.

-Antoine de Saint-Exupéry

Contents

Why Mars?

On September 19, 1783, before a crowd including King Louis XVI and Queen Marie Antoinette at the royal palace of Versailles, a duck, a rooster, and a sheep were lifted aloft to 1,500 feet in the first public demonstration of a flying machine carrying living creatures. The sheep was a stand-in for humans. The duck was used as an experimental control, because ducks are clearly accustomed to heights. As a flightless bird, the rooster was selected as a midpoint between terrestrial mammals and flying birds.

As with all new inventions and experiences, there were doubts and fears about sending earthbound creatures into the sky. What if they couldn't breathe? What if their hearts stopped beating? What if they died from sheer terror? As with most new experiences in the march of human progress, these fears proved groundless, and a new technological field was born[1].

When asked about the utility of the first flight, Benjamin Franklin responded: "What use is a newborn baby?[2]"

Humans evolved in East Africa, and subsequently expanded across the entire globe. However, we are by nature unsuited to many environments, with our feeble bodies and lack of fur. It is only by virtue of our technology that we can live in places as hostile as northern Europe, or

---

[1] In retrospect, these fears seem silly, especially since people had obviously climbed up the sides of tall mountains. However, I suppose they may not have understood how atmospheres work – it was 1783 after all.
[2] This is also attributed to Michael Faraday in an 1816 reference to electromagnetism, but Faraday clearly quoted Franklin, having made the reference 33 years later.

southern Patagonia. Were it not for this technology, humans would have no way to feed and provide for the billions of people on Earth.

Humans have an innate drive to explore. Perhaps it is a product of our nomadic history. As a species, we have spent far longer ranging over vast tracts of land in pursuit of migrating herds than we have in cities. It was by this means that humans arrived in Australia and North America[3].

We're constantly looking over the horizon for places to expand, explore, and settle. It's in our nature. After millions of years of evolution, we are just now at the point where we have the technology to settle other worlds and start the process of becoming a multi-planet species. But why should we bother? Since Earth is the most habitable planet we know of, why shouldn't we just stay here?

There are many answers to this question, but perhaps the simplest is basic survival. We already know of many threats to our civilization, ranging from nuclear war, disease, and resource depletion, to environmental collapse, and impacts from space. Considering how the list has been growing over the past 50 years, it seems likely that there are other threats we aren't even aware of. As Carl Sagan famously said: in the long run, every civilization must become spacefaring to ensure its very survival. A sustainable off-Earth presence would provide an insurance policy which could ensure our continued existence, and prevent us from going the way of the dinosaurs. Considering how little it would cost us[4], isn't it simply negligent not to take simple steps to avert potential extinction?

---

[3] Most likely around 40,000 and 14,000 years ago, respectively.
[4] As a species, we spend less than 0.1% of our global economic output on spaceflight. We spend more on pet food.

Although hardly a cause for alarm, considering the finite lifespan of the solar system, we must eventually travel to other stars[5]. Establishing settlements in our solar system, using existing and near-term technology, is the only way we are going to develop the advanced technology to one day take us to the stars. It was not by accident that the first human flight was in a hot air balloon made of paper. All technology is incremental.

Establishing branches of civilization on other worlds would not only insulate our species from disaster, but it might go a long way towards preventing it. As individuals, and as a society, challenge brings out the best in us. New cultures inevitably face new challenges, and develop novel solutions in response.

In human history, the most innovative and vibrant societies have always been those that remain outward looking. It is no accident that over the past hundred years, power has shifted from Europe to North America. The United States contains only 5% of the world population, but has produced over 20% of global wealth for the past 50 years, and generated over half of the world's inventions. Great endeavors inject optimism and virility into a society, which can have a profound impact on its fortunes. Even vicarious exploration has tremendous social utility.

At the start of the fifteenth century, China was the most powerful country in the world, supporting expeditions across the South Pacific and Indian Ocean, leaving a legacy from Singapore to the Middle East. China was vastly superior to Europe at the time, and was poised to be the dominant world power. China's naval technology dramatically outstripped Europe's: Chinese treasure ships were nine-masted monsters over 400 ft long[6], over six times the size of Columbus' largest ship, *Santa Maria*. However, fearful of the destabilizing potential of foreign influence, at the

---

[5] This is of course several billions of years off, and humans (provided we survive) will no longer be humans. Still, no need to procrastinate!
[6] About the size of a football field.

height of their power, the Ming emperors ceased all exploration after 1433. In subsequent decades, they banned all ship construction, overseas travel, and went so far as to depopulate the southern coastline for miles in order to cut all foreign contact. This disastrous policy left China moribund for centuries – until forced open by Europeans in the nineteenth century, and modernized very recently indeed[7].

Approximate size comparison of Zheng He's Chinese flagship and a contemporary Spanish ship

Another reason to go to Mars is the pursuit of scientific knowledge. There are older places on the surface of Mars than anywhere on Earth, and geological studies of Mars would reveal much about the early history

---

[7] This modernization has occurred at an alarming rate: in 1990, over 85% of Chinese lived in poverty. Today the number is less than 15%.

of the solar system. Studying impact history would tell us about current and future threats to Earth posed by asteroids and comets. It seems clear that Mars was once a much warmer and wetter place. Billions of years ago, Mars had a thicker atmosphere, and supported surface oceans and an Earth-like environment by virtue of a stronger greenhouse effect. What changed Mars into the cold desert we find today? Could the same thing happen to Earth? Can we one day restore Mars to its life-friendly past, and make it a true second home for humanity?

Comparative planetology is no mere matter of scientific curiosity. We learn a lot by studying the climates of Earth's nearest neighbours, without risking harm to this precious world. In the 1970s, atmospheric studies of Venus, a lead-melting inferno with crushing pressures, first alerted us to the dangers of a runaway greenhouse effect. Studies of planet-wide Martian dust storms alerted a Cold War Earth to the potential of nuclear winter – one more reason to stand back from the brink of mutual assured destruction.

Having humans on Mars would benefit our planet in other ways. Traveling to space and looking back on Earth has given us perspective as a species. There are no national boundaries from space. We've seen the thin blue atmosphere protecting us from the void. From Voyager, the most distant human creation, we looked back on Earth, a pale blue dot against the blackness of space[8].

These experiences underscore the importance of preserving our planet, the only oasis we currently have. The early environmental movement was fueled by this realization - in large part thanks to the astonishing Earthrise photograph, taken as Apollo 8 orbited the Moon on Christmas

---

[8] And more recently, NASA released an image of Earth rising beneath Saturn's rings, as seen from the Cassini spacecraft.

Eve, 1968[9]. Imagine the social impact of having humans gaze back on Earth from another planet for the first time. Imagine the unique perspectives that Martian settlers would have, struggling to survive on a hostile world. What would they be able to teach us?

Earth rising above the lunar horizon on Christmas Eve, 1968. Every human who has ever lived has inhabited this fragile oasis.

Human spaceflight and sustainability engineering are two sides of the same technology. The mere act of sustaining a small band of humans on another world would dramatically impact our water, energy, food production, and recycling technologies. Although many supplies would be sent from Earth, there would be huge incentives to produce locally. Supplying people on Mars is a logistical challenge not so different from supplying remote environments on Earth. Technologies like portable

---

[9] This photo was described by nature photographer Galen Rowell as the most influential environmental photograph of all time.

water filtration, efficient small-scale food production, and on-demand 3D printing would go a long way towards providing sustainably for all people on both Earth and Mars.

The fact that Mars supported oceans for ten times as long as it took Earth to develop life raises a very interesting question. Do all worlds with a friendly environment develop life? We know from recent planet-hunting missions like the Kepler Space Telescope that planets are extremely common. Most stars seem to possess planets, meaning that there are probably hundreds of billions of planets in our galaxy alone, with hundreds of billions of galaxies in the Universe.

Indeed, there are probably more planets in the Universe than all the grains of sand on all the beaches of Earth. Many of these seem to be about the right distance from the Sun to support liquid water[10]. So a question arises – do planets capable of supporting life develop life, or is Earth a rare phenomenon? Even signs of long-extinct microbes on Mars could indicate that life is abundant throughout the entire Universe, and that we are probably not alone[11].

Considering that Martian environmental changes took place over millions of years, that liquid water still exists under the surface, and that we find life even in the harshest and most isolated environments on Earth, it seems likely that pockets of Martian life may persist to this day. In fact, pieces of Mars have been blasted into space by impacts over the years, to be found as meteorites on Earth. Since some bacteria can survive in space, life on Earth may have originated on Mars, or vice-

---

[10] Known as the "Goldilocks" zone, after Goldilocks and the three bears – not too hot and not to cold.

[11] We are almost certainly not alone. Based on recent data from the Kepler space telescope, there are probably over 8 billion Earth-like planets in our galaxy (one of hundreds of billions of galaxies). This is more than enough to give everyone on Earth their own planet. I call dibs on Earth.

versa. If we found life on Mars, we might be able to learn a lot about the potential diversity of life by comparisons with Earth life.

Some scientists discredit the idea of sending humans to Mars, suggesting that scientific goals can be more inexpensively achieved by sending robots alone. While it is certainly true that robotic missions are much less expensive, they are also much less capable. As of July 2013, in almost a year on Mars, the $2.5 billion Curiosity rover[12] has traveled a single kilometer (0.6 miles). Opportunity, in almost ten years, has traveled almost 37 km (23 miles)[13]. Needless to say, a human on Mars could cover far more ground, far more quickly, and far more thoroughly. The time delay between Earth and Mars (between 4 and 24 minutes) means that every move must be planned out in advance with great precision, and decisions are made with great care.

Humans are also much better suited than robots for certain tasks, such as managing equipment to drill underground in search of water[14]. Moreover, having humans on Mars would not at all mean that we shouldn't use robots. Rather, it would mean that we could dramatically increase the efficiency of our robotic explorations. Human operators on the surface would cut the time delay to zero, and we'd be able to recover and repair damaged robots, or swap out their scientific instruments. Although we can answer some very specifically targeted scientific questions using robots, a human is still by far the most useful and versatile robot we could possibly send[15].

---

[12] A.K.A. Mars Science Laboratory.

[13] The off-Earth driving record goes to the Soviet *Lunokhod* 2, which covered around 42 km on the Moon.

[14] That's why they sent Bruce Willis to stop the asteroid in *Armageddon* of course!

[15] Planetary Geologist Ralph Harvey may have put it best: "If your goal is to answer very specific questions like how hard are the rocks on the surface of Mars, a robot is perfect. If you're questions are bigger, like, what is the history of Mars?, well, that's a hell of a lot of robots. But this could be done by just one

Investment in space technology invariably generates myriad spinoffs[16]. However, even more importantly, as Neil deGrasse Tyson points out, most major breakthroughs are indirect – that is, scientists accidentally make discoveries they aren't necessarily seeking. The entire field of antibiotics came about because Alexander Fleming accidently allowed fungus to grow on bacteria plates that should have been sterilized. Any investment into the cutting edge of technology is bound to have positive repercussions throughout society.

Our planet faces tremendous pressures from growing populations and diminishing resources. However, resource shortage is a matter of technology[17]. A single 2-km diameter asteroid could yield more iron and precious metals than we extract on the Earth in an entire year. Each hour, the Earth receives more solar energy than it would take to run our entire civilization for a year. Our world is swimming in water, so why do people go thirsty? Technological development is the best – perhaps the only – way to sustainably raise the standard of living of everyone on the planet.

Since the beginning, space research has been one of the most international fields of study, with more collaborations between scientists from around the world than any other field. Having people from various

---

or two humans because humans have intuition, where they've built a whole catalogue and can draw on it instantaneously."

[16] A partial list includes the digital computer, telemedicine, life support technologies, water filtration systems, cordless power tools, fireproof clothing, bulletproof vests, wireless data transfer, solar panels, insulin monitors, remote control, wireless switches, bomb detection devices, lightweight metal alloys, air traffic control systems, weather forecasting, and medical scanning technologies. Contrary to popular belief, NASA did not invent Tang, Velcro, or Teflon, although they do use them. NASA has a full list of these technologies on their website.

[17] During the Roman Empire, aluminum was more valuable than gold. Its inventor was killed by the Emperor to prevent him from sharing the state secret. However, aluminum is the most common metal in the Earth's crust, and with our extraction technologies, it is relatively inexpensive today.

nations and cultures working together to begin the settlement a new world would bring people together and help diffuse tensions. And though national pride is certainly a significant driver of space exploration, the international impact is clear. International surveys consistently rank the Moon landings as the United States' most significant achievement. There is little doubt that during the Cold War, they paid dividends in soft power across the globe[18].

A highly visible and international challenge like going to Mars would provide a positive vision for all of humanity. This kind of inspiration, though intangible, would give all people on Earth a shared goal and common purpose. It would be a beckon call to our youth to develop their technological skills, so that they too might participate in the greatest adventure of our age.

As astronaut Chris Hadfield has recently said while orbiting Earth: "Aim for the stars. You might not get exactly where you thought you'd be, but you will be doing things that suit you in a profession you believe in." Chris Hadfield's tenure of the International Space Station in early 2013 demonstrated that space still has the power to inspire. During his expedition, Hadfield performed science experiments with school children, exchanged tweets with Star Trek captains, shared the beauty of the Earth with a million Twitter followers, and serenaded us with songs from space. Plugged into today's technologies, with the power to reach the entire globe, we can only imagine the inspirational potential of a mission to Mars.

This kind of inspiration can have enormous benefits. In addition to the economic stimulus that derives naturally from employing the brightest

---

[18] Soft power is a term coined by Joseph Nye, Assistant Secretary of Defense for International Security Affairs during the Clinton Administration. It describes the ability to influence foreign nations through co-operation and emulation, instead of coercion or payoff - which are forms of "hard power".

minds on the cutting edge, there is a tremendous latent impact that can only be seen years later. In hindsight, even though people at the time questioned the value of sending people to the Moon, the Apollo program delivered a long-term economic stimulus that can still be traced today.

Interest in the Apollo program dramatically increased the number of science and engineering graduates at all levels. Every technological field experienced as surge as a result of the Apollo program - not just those with a direct connection. Children, whose first taste of science was Neil Armstrong's small step, grew up to initiate a tech revolution twenty years later.

Why not spend the money on Earth to do something like cure cancer instead? There is no doubt that there are other causes, some more important than space exploration. Clearly, sustaining human civilization on Earth is far more important than founding it on other worlds. However, this is a false dichotomy. Far from being in competition with the goal of preserving Earth civilization, establishing a human presence in space would significantly advance this goal.

Moreover, space exploration is not actually in competition with other critical tasks we face on Earth[19]. We spend a negligible fraction of our resources exploring space. Going to Mars wouldn't require much - if any - additional investment, if we approach it in the right way. In comparison to the several billions of dollars we spend on space per year (tens per

---

[19] The argument that we should "fix our problems on Earth before we go to space" is also silly because it will always be possible to find new problems to fix. What is the timeline for "fixing the problems on Earth"? 10 years? 15 years? Can you think of a time throughout history when we ever would have decided to explore? Just think of the social problems in Europe when Columbus set out. Indeed, to a large extent the explorations were motivated by troubles in Europe, as the Ottoman Empire choked off trade routes and threatened to overrun Christendom. Social problems in the United States seemed to climax during the 1960s. Looking back, the Apollo program seems like one of the only bright points of the decade – aside from the Beatles.

person), critical problems like repairing infrastructure, replacing fossil fuels, or fixing health care, social security, and education require trillions of dollars. Space exploration is one of the highest leverage fields we can pursue in terms of return on investment. The global cost of maintaining a small human settlement on Mars would be insignificant, but the rewards could be immense.

Although I support most scientific ventures, I think there is something fundamentally different about space. Unlike most scientific and social challenges, this one is not so open-ended. No new technologies are required. Unlike curing cancer, going to Mars is not actually a scientific challenge at all. It is an engineering challenge, more akin to an infrastructure megaproject than it is to basic research. This makes it fundamentally different from research programs, where investment may or may not yield results, and timelines are impossible to guess. Invest as much money as you want into cancer research, and nothing might come of it.

If we decide to go to Mars, despite potential setbacks, we can be fairly confident of success. This is one of the reasons that a Moon landing was selected as a target by Kennedy's advisors in 1961. It was a feat that, despite some uncertainty, was pretty widely accepted to be only a matter of incremental accomplishments – and we are much better prepared to send people to Mars today than we were in 1961 to send people to the Moon.

Why One-Way?

Assuming we should go to Mars, why should we go one-way? Though the idea admittedly sounds a little crazy at first, the more I think about it, the more I think that it is actually the best way to approach the humans to Mars challenge. Simply put, a one-way trip maximizes the return, minimizes the cost, and is by far the most logical way to organize a mission from almost every standpoint.

Although the technology to get to Mars essentially existed after Apollo[20] and NASA had originally planned to go to Mars as early as the 1980s, sending humans to Mars and bringing them home are two very different things. Since space is a vacuum without any drag, the fuel required to get to Mars isn't actually much higher than to get to the Moon.

In fact, since Mars has an atmosphere which can be used to slow an approaching spacecraft, the fuel required to land on Mars might actually be less than landing on the Moon. It is straightforward enough to get on a giant rocket on Earth and blast off to Mars. Similarly, to return, it wouldn't be too hard to get on a giant rocket on Mars, and blast off back to Earth. The problem is that there aren't currently any giant rockets on Mars, nor will there be for a long time.

This basic premise dictates why one-way to Mars makes so much sense. All NASA planning regarding human Mars missions has always assumed return missions with super-sized spaceships to be assembled in orbit

---

[20] Note that there is a big distinction between technology and hardware. When I say the technology exists to go to Mars, I don't mean that we have the actual spacecraft ready to go. I mean that we could start building them as soon as we decide that we want to invest in the mission, without any breakthroughs required.

using giant rockets that don't exist. These spaceships would then fly to Mars, explore from orbit, and then return to Earth. Not only would they need to carry their own mass plus the mass of landing craft, but they would need to carry fuel for the entire trip. This makes the mission dangerous, complicated, and - more importantly - expensive beyond our appetite for investment. As a result, the timeline for a human mission to Mars is still 20 years in the future, as it has continued to be for the past 40 years[21]. With no actual planning underway, this timeline will continue to advance indefinitely[22].

As soon as you decide that you are going to go one-way, everything becomes much simpler. It becomes a matter of sending people, landing, and sustaining the human presence through local resource production for bulk materials like water and oxygen, and resupply from Earth for other materials. Although there are certainly some extreme sacrifices required at the individual level for a one-way mission, for individuals with the right mindset, there could also be extreme rewards to be gained through unique experiences, rewarding challenges, and the knowledge of a life spent in the service of humanity.

Additionally, a one-way mission dramatically reduces the risks incurred by unproductive parts of the mission, and makes the overall mission much safer. The return trip is probably the most dangerous part, relying on equipment that will have been in space or on Mars for a long period of time, and will require doing something that has never been attempted:

---

[21] Actually, the timeline is slowly getting father away. Around 1970, NASA planned to land humans on Mars by 1986, only 16 years off. Now the NASA timeline seems to be the mid-2030s or later, over 22 years away.

[22] It's false to say that we don't know how to get to Mars and back. We do know how, and I have no doubt that if we were willing to invest the kind of resources into it that we did into, say, the Iraq War (well over $800 billion), we could do it in 10 years. The problem is that for whatever reason, we simply haven't been willing to invest these sums in space exploration. We may not need to if we approach the problem differently.

taking off from a different planet and returning to Earth. Assuming the Martian explorers know what they are getting into, can the additional risks incurred by a return trip even be justified when far more could be accomplished by staying?

We want to maximize the return of robotic missions - why wouldn't we want to do the same for human missions? This is not to say that humans are expendable like robots - but since the whole point of sending people to Mars is to have them living on Mars, solving challenges, it seems wasteful in the extreme to invest decades of effort and billions of dollars simply to return a few individuals to Earth. Wouldn't it be better if they stayed and continued to inspire, innovate, explore, and prepare for follow-up crews? Moreover, spending long periods of time in space incurs far more serious health effects than a stay on the Martian surface. Deciding to forgo the return trip halves most of these.

The idea of sending humans one-way to Mars is not new[23]. The "Mars to Stay" movement has been active for over 20 years. The earliest formal outline was presented by George Herbert at the Case for Mars VI Workshop in 1990. In a 2004 Op-Ed for the New York Times, Paul Davies argued that since "some people gleefully dice with death in the name of sport or adventure [and since] dangerous occupations that reduce life expectancy are commonplace, we ought not to find the risks involved in a Mars to Stay architecture unusual. A century ago, explorers set out to trek across Antarctica in the full knowledge that they could die

---

[23] Interestingly, both Russians and Americans briefly considered a one-way mission to the Moon in the 1960s, but neither intended for it to be suicide. The idea was to deliver at least a single human to the surface with sufficient supplies to survive, keep sending supplies, and then pick them up in a few years when the technological details were sorted out. Though it seems unlikely that such a scheme would have been implemented, one-way to Mars makes far more sense than to the Moon for a variety of reasons. It is interesting to speculate what might happened had Kennedy's goal been in jeopardy.

in the process, and that even if they succeeded, their health might be irreversibly harmed. Yet governments and scientific societies were willing sponsors of these enterprises. Why should it be different today?"

In another New York Times Opinion article, this one in 2009, physicist Lawrence Krauss asked "Why are we so interested in bringing the Mars astronauts home again? While the idea of sending astronauts aloft never to return may be jarring upon first hearing, the rationale for one-way trips into space has both historical and practical roots. Colonists and pilgrims seldom set off to the New World with the expectation of a return trip. If it sounds unrealistic to suggest that astronauts would be willing to leave home never to return, consider the results of several informal surveys I and several colleagues have conducted recently. One of my peers in Arizona recently accompanied a group of scientists and engineers from the Jet Propulsion Laboratory on a geological survey. He asked how many would be willing to go on a one-way mission into space. Every member of the group raised their hand."

Without a doubt there would be no problem finding volunteers, provided the parameters were well defined (i.e., planned settlement and not suicide). Many volunteers could even be found amongst the regular astronaut corps. Astronaut Bonnie Dunbar put it: "I've spent my life training to go to space. If it ended going to Mars, that's not a bad way to go". At the age of 76, the first woman in space, Valentina Tereshkova, has repeatedly volunteered for a one-way mission even if it were suicide, saying that it would be the best use of her life she could think of.

One-way to Mars could reduce the cost of a Mars mission by a factor of ten, could be done with NASA's current budget, and might be the only program that could revitalize the human space program. Moreover, since the manned space program is sold, at least philosophically, as a step towards colonizing other worlds, one-way to Mars seems to be a fairly self-evident fulfillment of that goal.

On a personal level, why would I volunteer for a one-way mission[24]? Partly, I don't consider the reasons why we as a species should go to be separable from the reasons why I would want to go: I would want to go because I think it would be the most important thing an individual could do with their life in the early 21st century. Partly, I don't think it would be honest to advocate doing something that I wouldn't volunteer for myself. And partly, I think it would be the greatest adventure someone could possibly have.

When people think about this sort of thing, they usually focus on the downside[25]. I think they often fail to consider the gains that would come along with the loss. You would be able to solve challenges on a daily basis, would feel like you were contributing to the advancement of humanity, and would be amongst the first humans to experience things like living on another planet or seeing a Martian sunset. Moreover, you wouldn't be alone, and you would be able to communicate, albeit with a time delay, with the entire home planet. I think that even on a personal level, the gains could well outweigh the losses[26].

And it may not actually be one way. Although I think that the purpose of a Mars mission is better served by the crew staying rather than turning around and coming home[27], perhaps there would be an opportunity to

---

[24] As an applicant for the Mars One mission, people ask me this a lot. You can find some of my interviews on my Youtube channel (AndrewRader).

[25] Our brains are wired to be very loss averse. In financial and social matters, we rate losses at least double as heavily as we do equivalent gains. This is of course illogical, but it is hardly the most illogical thing about the human brain.

[26] Although I admit to being overly optimistic at times, I don't think I'm unrealistic about the personal costs or living conditions. I expect that any such mission would obviously be very dangerous, not to mention extraordinarily uncomfortable at times.

[27] See Robert Zubrin's excellent book The Case for Mars or Google "Mars Direct" to see how this could be done. A Mars Direct architecture using fuel produced out of the Martian atmosphere would be an excellent way to return

return to Earth at some point. It's not necessarily a one-way forever mission. It could be, and you'd have to be ok with that possibility, but who knows what the future would hold?

---

from Mars, either on a planned return mission, or after a long stay as part of a one-way mission.

The Moon

Even among space enthusiasts who support the idea of human spaceflight beyond Earth orbit, there is a significant divide. Many ask why we should not go to the Moon first. The fundamental answer is that we already have.

We went to the Moon over 40 years ago, and by the time we go anywhere else, it will have been at least 50 years. Simply put, going to the Moon, while certainly interesting, will fail to inspire people. It's hard to get young people excited about doing the same thing their grandparents did. Though George W. Bush made a return to the Moon the primary goal of his 2004 Vision for Space Exploration, the Obama Administration found it easy to scrap the program shortly after taking office in 2008 because it lacked popular support.

Certainly returning to the Moon would be easier than going to Mars. However without a popular support base, it would always be at risk of cancellation. Setbacks, failures, and even catastrophes in pursuance of a Mars program might be taken in stride as they were for Apollo, but it is difficult to see how a major failure in a return to the Moon program would be perceived as anything short of incompetence.

The main advantages of a Moon base would be ease of supply, and rescue in case of emergency. There is also certainly some useful science to be performed – the Apollo missions explored only a tiny fraction of the surface and spent a total of around 24 hours conducting surface operations. It is also possible to do things like build a space telescope on the Lunar far side, where it would be protected from Earth interference[28].

However, the Moon is not really a useful dress rehearsal for Mars: the conditions on Mars are almost as different from the Moon as they are from Earth[29]. Nor is the Moon actually the great fuel depot it is often made out to be. Although it does harbor some water which could be converted to rocket fuel, this tends to be in difficult to access regions like the poles. And the fundamental problem with this argument is that you need to get the rockets to the Moon first, or assemble them at a lunar base. These are complex operations on Earth. At a barren Moon base lacking infrastructure, they would be unfathomable. Ultimately, it is easier to journey to Mars directly from Earth.

The Moon also makes a poor choice because it lacks the potential of Mars. Mars possesses many Earth-like features, with almost as much surface as all the land areas of Earth[30]. Just like Earth, Mars experiences seasons, with a year length of about two Earth years long[31]. A day on Mars is just slightly longer than a day on Earth, at 24 hours and 39 minutes. Critically, this means that plants could be grown in a Martian greenhouse using natural light. By comparison to Mars, the Moon is

---

[28] I also think there is a good argument to be made that the Moon has an advantage over Mars simply because it is more visible in the sky. The inspirational value of being able to look up from Earth and say "people are living up there" would be huge. You can do the same for Mars of course, but Mars tends to blend in a lot better with the stars. How many of you can reliably identify Mars without a Google sky map?

[29] An asteroid mission might be a useful precursor to going to Mars. Mars' moons, Phobos and Deimos are not very useful in my opinion, because going to them carries most of the risks associated with going to Mars (even more if you count the human factors associated with being in space), and a lot less of the rewards.

[30] I.e., not counting oceans, Mars' surface area is 93% of Earth's.

[31] In fact, the seasons are slightly more pronounced on Mars, with an axial tilt of 25° as compared with Earth's 23.5°.

essentially a dead rock in space. The Moon has no atmosphere, less than half the gravity[32], and fewer resources of any kind.

Mars is the only other world that we have the technology to reach today which possesses the full spectrum of resources necessary to support sustainable human settlement. Mars is the fourth planet from the Sun, at the edge of the habitable zone, about 50% farther than Earth. This means that Mars is cold, but not so cold that humans couldn't live in heated structures and safely explore using pressure suits.

At the equator, Mars receives about as much sunlight as Scandinavia. With an equatorial temperature between +20° C (68° F) and -70° C (-94° F), Mars is cold. However, the thin air means that far less heat would be lost compared with similar temperatures on Earth[33]. This effect is similar to the difference between water and air temperature – since water takes away heat much faster, it feels much hotter or colder than air at the same temperature. Had the Titanic passengers been plunged into cold air instead of icy water, they would have survived for hours or days instead of minutes.

Perhaps most importantly, Mars has abundant sources of water[34]. Although the low temperatures and pressures mean that liquid water

---

[32] Lunar gravity and Martian gravity are around 17% and 38% of Earth's, respectively. Since bone and muscle loss is an adaptation to the gravitational environment, this would likely make a huge difference in the long term health of human settlers, and might have other impacts like plant growth. Bottom line: the more Earth-like, the better.
[33] Similarly, even though space is very cold, at only a few degrees above absolute zero,
(-273° C), overheating is often as much a concern as freezing, because it is much harder to dissipate heat to a vacuum.
[34] Estimates vary on how much water, but much of the terrain seems to be permafrost containing around 6% water. If all the water on Mars were melted on a perfectly flat world, the sea level would be between 100 and 1500 ft, with an average estimate of around 600 ft. This compares with an average Earth sea level of around 6000 ft.

cannot long persist on the surface, our robotic explorers see evidence of frequent releases of underground liquid water. The Martian atmosphere is composed mainly of carbon dioxide, is useful for growing plants and can also be converted into oxygen and methane – perfect for rocket fuel or use as an auxiliary power source. While most supplies can be sent from Earth, shipping large quantities of water and oxygen as far as Mars would be impractical. If we want to settle the solar system, the starting point is obvious: Mars.

The Chicken and Egg of Technology

Assuming we should go to Mars, why should we go now? Currently there is no overriding goal driving the human spaceflight program. NASA today is often regarded as a jobs program, or a way to gain political favor in Florida and Texas. There is no clear directive from the President or Congress. The result is a program that shuffles along with little incentive to take risks or deliver results[35]. Sadly, NASA has largely devolved into a bureaucratic organization with too many disparate views and objectives.

Wernher Von Braun recognized the utility of maintaining a smaller, more efficient organization during the Moon race: "If we'd been more people at NASA, we'd have failed"[36]. There are certainly notable exceptions: the robotic space exploration program, particularly the recent Curiosity rover, are shining examples of what NASA can still do. However, NASA fundamentally lacks direction, and most of the fault lies with politicians.

It's not even primarily a matter of funding. The human space program needs to have a coherent and specific plan, directed by political leadership, in order to function efficiently. The attempt by George W.

---

[35] Do you work better with a goal and deadline? You're a procrastinator, aren't you? The same thing is true of large organizations, including the government (gasp!), and even NASA.

[36] Another great quote from Von Braun, who originally designed the V2 rocket for the Nazis during the Second World War: "I aimed for the Moon, but hit London instead".

Bush to provide this with the Vision for Space Exploration in 2004 was cancelled by the Obama Administration[37].

The "goal", if you can call it that, is now a vague statement about going to asteroids and Mars in 15-25 years. Unfortunately, a political goal lasting much longer than a presidential term is practically worthless. If Kennedy had promised to go to the Moon by 1980 rather than by the end of the decade, it is questionable whether we'd even be there today[38]. Any program that fails to make significant progress during one presidential term is likely to be cancelled during the next.

The future isn't just going to happen if we wait around for it. Without incentives, technological development stagnates. The technology to go to Mars was conceived in the 1950s, and essentially existed in 1970, yet almost 45 years later we are no closer to the goal. It is true that the space race was a product of the cold war, so perhaps we space enthusiasts are overly optimistic in expecting a concerted national effort. However, shouldn't we all demand an efficient program that delivers concrete results?

Technology follows purpose, not the other way around. NASA's Mars plans usually propose futuristic spaceships with orbital assembly, in-flight refueling, advanced radiation shielding, artificial gravity, and the capability to spend years in space. However, we simply can't wait for hyper-advanced propulsion, let alone warp drive, because these won't be

---

[37] The Obama Administration as a whole has been rather unsupportive of space exploration, even proposing cuts to the highly successful Mars robotic missions. However, as this is being written, the Republican controlled House is proposing even deeper cuts to NASA.

[38] One of the first things Nixon did after the first Moon landing was cancel the remaining Apollo missions and scrap the Saturn V rocket – the largest and finest rocket the world has ever seen. If you want to know why we aren't on Mars yet, you might want to ask Richard Nixon.

invented unless we start by pushing ahead with the technologies that we already have.

Necessity is the mother of invention[39], and major feats can be accomplished with limited technology. Early European explorations were performed using flimsy coastal ships[40] because robust sea-going vessels hadn't been invented yet – and never would have been without the incentives generated by global settlement and commerce.

Once human bases are established beyond Earth orbit, there will be enormous incentives to both improve our spaceflight technologies, and dramatically reduce their costs. Space travel is expensive, but this is largely due to the high costs associated with breaking free of Earth's gravity[41]. Commercial incentives will drive technologies at a faster pace, and space travel will gradually become cheaper and easier. Owing to its uncomfortably high gravity, Earth may eventually be a virtual no-fly zone in a commercial network between Mars, the Moon, and asteroids[42]. We need to be bolder and less risk averse when we think about going to space.

---

[39] And procrastination is the mother of necessity, which is one reason we find it difficult to take action on major issues until they become critical.

[40] Although compared with the Viking ships that made trans-Atlantic crossings hundreds of years earlier, they were luxury liners. You can do a lot with limited technology.

[41] If you lived at an asteroid base, you'd have to be very careful not to launch yourself into space every time you took a step.

[42] My favourite asteroid (now "dwarf planet") is Ceres, the largest in the asteroid belt. It's essentially a giant low-gravity ice cube between Mars and Jupiter, which could eventually provide propellant for journeys to the outer solar system.

Trajectory of a Space Program

NASA was founded as a direct response to Sputnik[43]. Had the Soviets not been the first to get to space during the Cold War, the space race and Moon landings might never have occurred, and certainly not when they did. From its formation in 1958 until July 20, 1969, NASA had a clear purpose: to beat the Soviets in space. Although there had always been space enthusiasts and dreamers, never before had they been given the resources to vigorously pursue space research or activities[44]. Although most prior space research had been military, NASA was formed specifically as a civilian organization to assure the world that US intentions were peaceful[45].

This clear sense of purpose provided and unified vision helped NASA perform at its best. NASA was infused with additional funding, peaking at 4.4% of the federal budget in 1966. As a percentage of spending, NASA's funding has been declining slowly since the 1970s, and now stands at around 0.5% - half a penny on the dollar. NASA seems to linger on life support because, while it would be political suicide to renounce the achievements of Apollo by shutting it down, it would be equally damaging to spend vast sums on a program that doesn't seem to deliver concrete results to the American public.

---

[43] It was attached to NASA, which had been the equivalent for Aviation only since 1915.

[44] With the exception of military purposes, particularly in Germany during the Second World War.

[45] There was even a committee to decide if planting a US flag on the Moon was too nationalistic: it was decided that this impact was offset by the plaque "we come in peace for all mankind".

In some technological sense, we have regressed since 1970. We no longer have the capability to send people to the Moon. The Saturn V rocket was far more capable in terms of lifting payloads than anything we have had since. The human spaceflight program since the 1970s has focused on the Space Shuttle and Space Station operations, partially because they prepare us for longer duration missions, but mainly because political constraints haven't allowed for a bolder vision.

Around the time of the Moon Landings, NASA came up with a list of programs to build on their success. These programs included a permanent human presence on the Moon, an Earth-orbiting space station, flyby missions to asteroids and Venus, a reusable space plane, and a mission to Mars. All of these were scheduled to be well underway by the 1980s, with a manned Mars landing planned before 1990. However, with the Cold War space race won, the government saw little purpose in funding space. NASA, facing impending cuts, decided to pursue just a single project. The space plane was selected because it promised to dramatically reduce the cost of access to space. It was hoped that this would open the way to more ambitious projects in the future.

Sadly, although the Space Shuttle was a technological marvel, owing to early design compromises[46] the program was a massive strategic failure. Instead of reducing the cost of space access, it dramatically increased it. In stark contrast to the Saturn V Moon rocket, the Space Shuttle was by far the most expensive launcher ever operated. There are various ways to count cost, including cost per flight, total program cost, or average cost over a given period of time. The Space Shuttle comes up short in all of them. In current dollars, the entire cost of the Saturn V program was

---

[46] Specifically, the shuttle was i) only partly reusable and as a result expensive and dangerous, ii) combined payload and crew – the last thing you want to do for a space vehicle, iii) too large and complex as a result of requirements imposed by the military as part of a space collaboration program that was later abandoned, and iv) never achieved anywhere near the launch rate originally promised.

between $3 and $6 billion per flight, for 13 flights. However, the individual production cost was much lower, at around $565 million[47]. Even factoring in a couple hundred million for assembly and launch costs, at 125 tonnes to low-Earth orbit, this works out to around $6000/kg.

The development costs for the Space Shuttle were probably higher than for the Saturn V, but over ten times as many flights were flown. The cost per shuttle flight has been estimated between $200 and $450 million[48]. With a capacity to orbit of less than 25 tonnes (1/5th of the Saturn V), cost to orbit was double as high, at something like $12,000/kg.

Even if we take the entire cost of the Space Shuttle program (over $190 billion), and divide it by the number of flights (135), we get around $1.4 billion per flight - still much more per kilogram than the high estimate for the Saturn V. While it is true that the Saturn V was just a launcher, and the Space Shuttle could repair and theoretically retrieve satellites, this capability was rarely used – mainly because it would usually be cheaper to simply build a replacement satellite instead[49].

The problem with the Space Shuttle was not only the initial design compromises, but the fundamental fact that, for the launch volumes associated with space flight, simplicity and assembly line production provide far more cost savings than reusability. This raises the question of whether the US really won the space race after all?

---

[47] "Stages to Saturn", Roger. E. Bilstein. NASA History Series, 1996.
[48] "The Rise and Fall of the Space Shuttle", Pat Duggins. University Press of Florida, 2007.
[49] The notable exceptions to this were the wonderful Hubble Space Telescope repair missions.

While the Russians retained their efficient and reliable vehicle, the US scraped its own. NASA ended up with an expensive[50] vehicle that had to be inspected by armies of workers between flights. By contrast, the Russians stuck to the Apollo model, and continued to rely on the disposable rockets and capsules that now carry American astronauts into space[51]. Since the 1960s, Russia has maintained the highest launch rate of any country, typically launching twice as many rockets as the United States[52].

The shuttle program did perform some useful science, and carried over 500 people to space. However, accomplishment simply can't be measured in terms of people carried to orbit – in fact, the high volume of launches actually has a negative consequence on public opinion, making space travel seem routine when it is anything but. Fewer, more impactful missions would provide much more utility, and at a much lower cost. Moreover, the Skylab and Mir programs proved that such science can be performed using Apollo-style hardware.

In the second half of its life, the Space Shuttle was almost entirely devoted to assembling the International Space Station (ISS). Although the ISS has again yielded valuable scientific data and operational experience[53], it is doubtful whether these can justify the cost of the program. The most common reason stated for the existence of the space station is to learn how to live in space while still being close to Earth.

---

[50] And also dangerous as it turned out, with the loss of 2 shuttles and 14 crew, as compared with no deaths on the Russian Soyuz since 1971.

[51] Actually, the Soviets did build a copy of the Shuttle called Buran. They flew it once in 1988, and then scraped it. Maybe they decided it wasn't worth the effort. The USSR had a penchant for copying US technology: when a B-29 bomber crash-landed there in 1945, they reverse engineered it bolt-by-bolt and built hundreds of copies. But they probably thought it was unnecessarily complex too...

[52] Of course, this is also because Russian satellites are far less reliable than US ones, so they have to be replaced more often.

[53] Particularly in the areas of life support, space biology, and telemedicine.

NASA usually flies missions to ISS that are around 6 months long, and plans to increase these to a year. However, several missions longer than a year had already been performed on the Russian Mir even before ISS assembly began.

In some ways, this rationale for the space station effectively amounts to subjecting astronauts to unnecessary negative health effects just to discover how to best deal with them. How far removed is this from purposely infecting people with a disease in order to test potential cures? If we are going to send humans to space at great cost and risk, wouldn't it be better to pursue a broader mission at the same time?

In reality, the primary reasons for the space station program were i) political, as a means of employing Russian scientists who might otherwise seek work designing weapons after the Cold War, ii) diplomatic, enlisting the co-operation of allied nations[54], iii) to justify the shuttle program, and iv) simply to give NASA something to do. Although it is true that the robotic arm[55] and cargo bay made the shuttle perfect for assembly and repair work, it is hard to see how the same thing couldn't have been accomplished with the Saturn V. Indeed, with more lifting capacity, the Saturn V could have carried all ISS components to orbit on fewer than ten launches.

Even many scientists, both inside and outside the space community, have questioned whether the 15 years and over $150 billion invested in the space station program have been well spent[56]. Lord Rees, British

---

[54] Although international co-operation can be a great thing, it can also be detrimental to progress, blunting competition and reducing the justification for national achievement.

[55] In Canada, we are justifiably very proud of the Canadarm and its ISS successor; however I've noticed that most Americans prefer to call it the "Remote Manipulator System (RMS)". Come on America, you even named a type of cheese after yourself!

[56] ISS does have some high profile supporters: Chris Hadfield told BBC News in 2013 that it is useful "because we can at any moment, when we have made a

Astronomer Royal, opined that: "No one would regard the science on the space station as being able to justify more than a fraction of its overall cost. Its main purpose was to keep the manned space program alive, and to learn how humans can live and work in space. And here again the most positive development in this area has been the advent of private companies which can develop technology and rockets more cheaply than NASA and its traditional contractors have done."

So perhaps, by creating an unnecessarily expensive space infrastructure, the space station's main benefit has been to stimulate the development of a more innovative, less expensive way to approach space exploration.

---

stupid mistake with a design, or an emergency that we hadn't recognized, or because of human health, get in our spaceship and come home. We are learning how to live in space and so ISS probably will be a stepping stone to Mars". I don't really disagree with these statements – I'm simply not sure that these benefits are worth the delay and cost. I will discuss this further in the chapter on risk.

Private Space

In the long run, space needs to pay for itself. However, large initial government investments in new transportation technologies are the norm rather than the exception. Columbus' expeditions were government-funded, as were the voyages of Hudson, Magellan, and most other early explorers. Most historical colonization efforts, like the settlement of Jamestown, were government ventures. The European outposts established throughout the world have had a dramatic impact on the history of the world, and were all paid for by government entities.

In the history of transportation, we see a consistent pattern. Canals, sailing ships, steamships, the railroad, motor vehicles, and even airplanes were invented privately, but then developed to a commercially viable level by government investment. Historically, this has often been through the military during wartime, but it doesn't have to be. The East India and Hudson Bay companies were ostensibly private stock companies, but neither would have survived without Royal support and protection.

There are many ventures which have the potential to yield significant returns in the long run, but whose high initial costs make them unattractive to private investors. One of the primary roles of government is to promote long-term prosperity, even when returns are beyond the investment horizon of the private sector. However, the relative immunity from financial accountability is precisely why the government also tends to be horribly inefficient as compared with private industry. The government builds the railroads, and private industry takes it from there.

Fortunately, we are now seeing the first signs that space can in fact yield a profit, and also that private industry is capable of participating in the space sector while dramatically reducing costs[57]. Since 2004, when SpaceShipOne became the first commercial vehicle to reach space, we've seen over a dozen private companies join the private space race[58].

The nearest on the horizon, and most likely to make a profit in the short term, are the suborbital tourism companies. Of these, Virgin Galactic will likely be the first out of the gate, with its first commercial flight planned before Christmas this year (2013). Others like XCOR are hot on their trail to take advantage of the emerging market. These flights are short hops into space, not fast enough to stay in orbit[59].

Imagine firing a cannon ball in an arc, faster and faster. Suborbital flights are like cannon shots that go into space and then come back down to Earth. By contrast, the trajectory of the International Space Station is an orbit – meaning that it is always falling towards the Earth, but is going so fast that by the time it would hit, the Earth is no longer underneath. The result is that it just keeps going around in circles. Suborbital flight requires a lot less fuel, and is therefore a lot easier to achieve. This makes it relatively inexpensive and therefore marketable to the rich for a couple hundred thousand dollars. Orbital flight is much harder, and only a small number of nations and companies have been able to achieve this.

---

[57] It is not always clear what people mean when they say "private" or "commercial" space. In one sense, we've had private space for decades – most military and NASA contractors have, after all, been private companies. The main distinction here is that "new space" companies seek profit outside the government sector, and as such are driven by a desire to perform while making serious efforts to cut costs.

[58] A list of more prominent companies in no particular order include SpaceX, Virgin Galactic, XCOR, Bigelow Aerospace, Blue Origin, Sierra Nevada, Orbital Sciences, Starchaser Industries, Deep Space Industries, and Planetary Resources.

[59] The first two US space flights during the Mercury program were similar suborbital flights, including the flight by Alan Shepard, the first American human in space. Although around a dozen monkeys and a chimpanzee flew before him, so he wasn't the first American primate in space.

The most successful new company in the orbital space arena is SpaceX, an entirely private company founded by Paypal multi-millionaire Elon Musk. Their business model aims to maintain a small but dedicated workforce to minimize costs. Falcon 1, the first vehicle fielded by SpaceX, was the first privately funded launch vehicle to reach orbit when it did so in 2008[60]. SpaceX quickly moved on to the much larger Falcon 9[61], which has had 4 fully successful flights to date since 2010, with one partial failure – quite a good track record for such a new rocket.

Under the NASA Commercial Orbital Transportation Services (COTS) [62] program, SpaceX has now flown two resupply missions to the International Space Station. While still reliant on NASA funding, it looks like SpaceX has so far succeeded in its goal of reducing the cost to orbit. Compared to the shuttle launch price of several hundred million dollars, a Falcon 9 heavy is priced at around $55 million, and is capable of lifting about 10 tonnes to orbit. This gives a cost of around $5,500/kg to orbit, or slightly less than the Saturn V, and less than half the shuttle price.

The SpaceX Falcon 9 version 1.1, first flown in 2013, performs significantly better, at less than $4,000/kg. Even more exciting is their Falcon Heavy launch vehicle, scheduled to fly in 2014 or 2015. If SpaceX is able to meet their target, Falcon Heavy will be the largest operational rocket in the world, capable of lifting up to 53 tonnes to orbit, at less than $2,500/kg – 5 times less than the Space Shuttle. Over the next few years, we will see if this can be done.

---

[60] Another contender is Pegasus, but this is quite small and must be launched from an aircraft.
[61] They are named after the number of rocket nozzles: Falcon 1 has 1, Falcon 9 has 9.
[62] The space industry loves acronyms, especially TLAs (3-letter acronyms). For example, "Maximum Absorbency Garments" (MAGs), A.K.A. space diapers.

Not only are unit costs much lower[63], but development timelines and costs have been dramatically reduced. Total costs for Falcon 9 development were around $300 million, less than a tenth of what NASA would have spent. With over 25 commercial launch contracts, SpaceX is forcing other aerospace entities around the world to develop their own competitive low cost solutions[64]. Similarly, the SpaceX Dragon crew capsule was developed for around $300 million over 4 years, well under half the development time and cost of the NASA/Lockheed Orion[65]. The Dragon capsule is currently being used for resupply missions to ISS, and is planned to fly missions with crew starting after 2015. In a very real sense, SpaceX is in the process of making space more affordable. Yet the best may be to come. Although Elon Musk is obviously an able businessman, his primary goal is to change the world by getting humans to Mars.

Three other space companies deserve special mention here. The first is Paragon Aerospace, a world supplier of life support systems, having built hardware for over 70 space missions[66]. With expertise in water recycling and closed ecological systems, Paragon would be a prime candidate for any long duration mission. They are an integral part of Inspiration Mars[67] and Mars One Planning. The second is ILC Dover, best known for

---

[63] Hopefully declining further as production volumes increase and partial reusability is introduced.

[64] Other American companies, in addition to the Europeans and Russians, are all attempting to develop low cost launchers, largely spurred on by this competition.

[65] This capsule was conceived as part of the George W. Bush Vision for Space Exploration, and has been known as "Orion", "Crew Exploration Vehicle" (CEV), and is now known as the "Multipurpose Crew Vehicle" (MPCV). I'll call it Orion here for historical reasons – and because, seriously guys, it's just a better name.

[66] Keeping with the unnecessarily complicated terms in the space industry, life support systems are called "Environmental Control and Life Support Systems" (ECLSS).

[67] Inspiration Mars is the planned 2018 Mars Flyby organized by multi-millionaire Dennis Tito.

building the spacesuits that have outfitted every American astronaut since Apollo[68]. They would build Mars spacesuits, and potentially inflatable habitation modules as well. The third is Bigelow Aerospace, founded by Hotel magnate Robert Bigelow. Naturally, Bigelow now plans to build space hotels using inflatable habitation composed of several layers of Vectran, a composite material twice as strong as Kevlar[69].

This approach offers the possibility of launching very large enclosed habitations into space, making space modules that are roomier, lighter, and cheaper to launch. Bigelow launched two demonstration models into space in 2006 and 2007, is now building an inflatable module for ISS – the Bigelow Expandable Activity Module (BEAM). Not only would such inflatable habitations be useful for space stations and voyages to Mars, but they could be an excellent way to make roomier Mars bases at a reasonable cost.

---

[68] The Russians have their own spacesuits, including the unpopular "penguin suit" used to generate tension with springs so that cosmonauts automatically exercise while going about their daily routine. I don't know what Chinese Taikonauts wear – go look it up!

[69] In case you didn't know, Kevlar is a strong material used in sporting gear, bullet proof vests, etc.

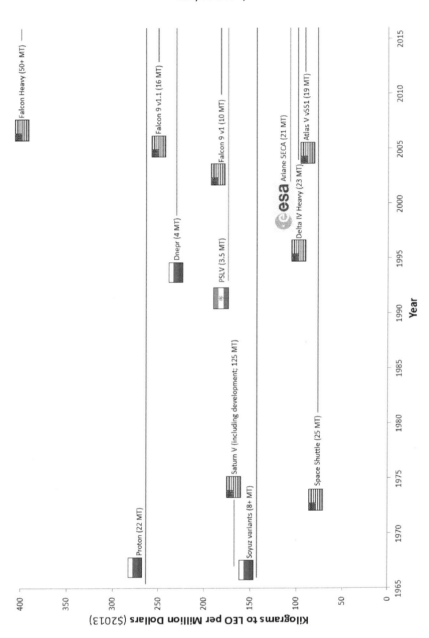

Comparison of launch systems: approximate payload per dollar to low-Earth orbit (LEO)

Getting to Mars and Staying There

Wernher von Braun spent a lot of time thinking about how to get to Mars. He first outlined the technical specifications for a mission in his 1952 book *Das Marsprojekt*. Launching in 1965, his mission called for ten spacecraft with over 70 crewmen on a several-year mission, spending 443 days on the Martian surface before their return to Earth. Von Braun had run the Nazi rocket program, because they provided the funding he required to pursue his space dreams. He compared a mission to Mars to a military operation, asserting that "the logistical difficulties of a mission to Mars are no more challenging than a minor military operation operating over a limited theater of war"[70]. Most planning since, though far less ambitious than von Braun's, has maintained the same approach: large, redundant, and robust, using technology to conquer challenges.

Considering that ISS took more than 31 launches over a decade to complete, assembling this kind of vehicle would require a very long time. Using existing rockets, NASA's Exploration Missions and Systems Office estimated that it would take 70-80 launches to assemble such a Mars spacecraft. There are also numerous problems with large-scale orbital assembly. Since all hardware would be designed and built on Earth over a number of years, if any launches fail, the whole mission

---

[70] Military space is still larger than civilian space around the world. The Chinese space program is highly integrated with their military, and the US military has space program entirely separate for NASA. In fact, the US military space program has a larger budget than NASA. I often fancifully dream about what civil space could do if it were as well funded as the military. In fact, NASA spends less than 5% as much, even though public opinion polls show that most Americans think their budgets are approximately equal.

timeline would be utterly ruined[71]. Further problems include the fact that you don't want hardware sitting in space longer than necessary, and cryogenic fuel starts to boil off into space as soon as it is launched[72].

Assembling a large vehicle in Earth orbit was precisely what NASA had originally planned to do to get to the Moon. However, the idea was eventually replaced by the much lighter and cheaper lunar orbit rendezvous. With this approach, a single launch of a Saturn V carried a command vehicle to lunar orbit. From there, a lightweight disposable lander carried two explorers to the surface for a short series of excursions[73]. Considering that the Apollo program was already suffering time and cost overruns, a NASA fixated on a massive orbital construction project would never have made Kennedy's deadline - if they ever got to the Moon at all.

Although I would love to see NASA funding doubled and maintained at a penny on the dollar as Neil deGrasse Tyson advocates, for political reasons we space enthusiasts simply can't count on a massive influx of funding to support a mission to Mars. With constrained budgets, we can't afford to be wasteful. If we want to do more, architectures must be as lean as possible.

---

[71] One solution would be to build two copies of everything. I think this would be a fantastic approach to every mission, including robotic exploration missions, considering that the majority of investment goes into R&D rather than actual production.

[72] This, along with the difficulties of in-space fuel transfer, is one of the main issues with using cryogenic fuel depots. Liquid hydrogen and liquid oxygen boil off at approximately 4% and 0.5% per month, respectively.

[73] Interestingly, and contrary to what the world believed during the Cold War, the Soviets were secretly participating in the Moon race. However, driven by factionalism, they pursued two different and competing Moon programs: one to flyby the Moon, and a second to land a single Cosmonaut. The Soviet participation in the Moon race was discovered in the early 1990s when American scientists accidentally ran into a Soviet lunar lander in a Moscow museum. It is interesting to speculate what might have happened had the Soviets concentrated their resources on a single lunar program.

The Minimalist Approach

In 1725, the Danish explorer Vitus Bering led several hundred men on a 4,200 mile (6,700 km) overland journey[74] from the Russian capital of St. Petersburg to the far Russian East of the Kamchatka peninsula[75]. Following rivers and natural routes, it was a perilous journey taking over a year to reach the Pacific Ocean, and losing many men along the way. However, this was only the beginning of the endeavor.

When they got to the Pacific Ocean, they had to found a town, build their ships, and then set off on their real objective: a sea voyage to explore and claim the Northern Pacific, and expand Russian influence into North America. This they succeeded in doing, with two sea expeditions along the coast of Alaska. Later Russian explorers pushed further, establishing Russian outposts as far down the Pacific coast as California[76].

---

[74] This represents approximately double the distance covered by America's Lewis and Clark expedition (2,100 miles), and may be the longest overland expedition of all time. Combined with the subsequent sea voyages, I would nominate the Bering expeditions as the most challenging exploration in history.
[75] The history of Russian expansion into Siberia is both fascinating and poorly publicized. Starting in the 1560s, they explored and expanded into double the area of the United States. Moreover, they did it in less time, and 200 years earlier. This was, without a doubt, the most impressive permanent national colonization exercise of all time. Russia reached the Pacific Ocean in less than a hundred years, and was the first European nation to establish formal relations with China, with the Treaty of Nerchinsk in 1689.
[76] Russia later sold Alaska to the United States in 1867 for $7.2 million dollars (around $20 billion today). Ridiculed at the time, this seems in retrospect to be one of the greatest land deals in history, along with the Louisiana purchase, and acquisition of the island of Manhattan from Indian tribes for $24. British representatives declined to purchase Alaska, which still irks me because it would then have been part of Canada where I come from. I also wonder about how different the Cold War would have been with Russian bases down the Pacific Coast within 500 miles of Seattle. On a further unrelated topic, the US

Bering's expedition, along with most expeditions undertaken during the age of exploration and subsequent polar expeditions, would never have been possible had the explorers not relied on local resources. Lewis and Clark didn't bring enough food for the trip; they brought hunting rifles[77]. Moreover, in comparison with space travel, freight costs were proportionally much lower during age of exploration. We need to think more like these explorers if we want to be serious about sustainable spaceflight. Although it will be decades before we can think about manufacturing high tech equipment in space, bulk items like fuel, oxygen, water, and to some extent food, must be acquired in space – and Mars is the best place to do this.

All of this means that we need to focus on near term, minimalistic missions. A one-way mission provides the most realistic near-term architecture possible, and could be sustained with current spending. As we do with the International Space Station, we will have to send regular supply flights. These will have to be planned even more carefully, because with a two-year launch window to Mars, we'll need to ensure that nothing is missed[78]. How would such a mission look?

There are at least two possibilities. One possibility is that SpaceX is successful with their planned Falcon Heavy development and test flights. Although the Falcon Heavy would be more capable than any vehicles currently in operation, it still has less than half the lifting capacity of the Saturn V. With an improved upper stage, a Falcon Heavy would be able

---

president to add the most territory to the United States was not Thomas Jefferson with the Louisiana purchase, but James K. Polk with the Mexican War and multiple land deals with Britain. I like history.

[77] Some people think rifles didn't exist before the Crimean war. This is false: rifles did exist, they just took far longer to load than the far less accurate muskets, and so were only used by hunters and military sharpshooters.

[78] It would of course also be highly advisable to send at least two, and hopefully three, sets of all supplies for each resupply mission to add redundancy and account for accidental loss of supply vehicles enroute.

to land over 10 tonnes on the Martian surface in a Dragon capsule[79]. Dragon has been designed from the start with the capability of landing on Mars. Even though Dragon can probably be upscaled from a diameter of 3.8 m to around 5 m, it is still very small. Thus, inflatable habitation would almost certainly have to be used on both the trip to Mars, and for the base on the surface. Moreover, this kind of architecture would be barebones in terms of supplies - alarmingly so if it were to carry 4 crew as planned. Several launches would likely be required, with a small Mars transfer vehicle assembled in Earth orbit[80].

Another possibility is that NASA's Space Launch System (SLS) is completed on schedule by 2017, along with the NASA/Lockheed Orion spacecraft planned for first launch in 2014, and first manned flight around 2020. With approximately the same lifting capacity as the Saturn V, the Space Launch System[81] would be more than capable of launching Orion and enough supplies to Mars for this kind of mission. However, this would depend on NASA continuing on their current path, which given the political climate is certainly not a sure bet. Additionally, even if these vehicles were completed on schedule, while using them for this kind of mission would be technically possible, it is difficult to see how NASA would ever support such a mission with no provision for return[82].

---

[79] Minus the weight of the Dragon itself, which is around 4.2 tonnes.

[80] Payload-wise, it might be possible to use other rockets like the Russian Proton, but it is difficult to see how this would be practical with mixed hardware on a short schedule.

[81] I wonder when they are going to give it a real name other than the TLA it currently has. All the good rocket names seem to be already taken.

[82] It is even a bit difficult to see how private industry would support this kind of mission, but this is at least somewhat less far-fetched.

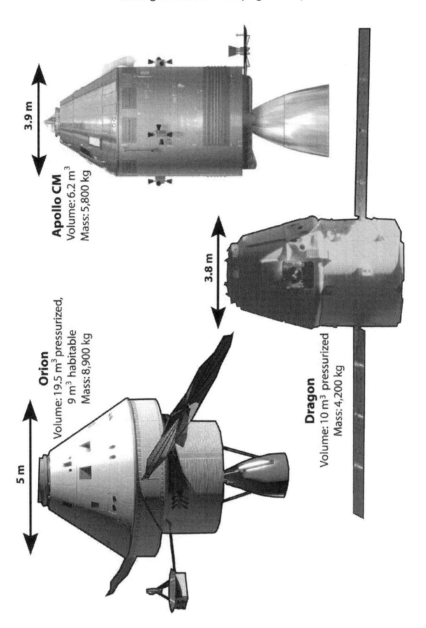

Apollo CM
Volume: 6.2 m³
Mass: 5,800 kg

Orion
Volume: 19.5 m³ pressurized,
9 m³ habitable
Mass: 8,900 kg

Dragon
Volume: 10 m³ pressurized
Mass: 4,200 kg

3.9 m

3.8 m

5 m

Scaled comparison of crew capsules. Note that both Dragon and Orion
would require larger service modules for a Mars mission

Landing on Mars

Mars is a strange environment, with atmospheric pressures about 1% of those on Earth[83]. This makes Mars vastly different from deep space or the Moon, but also vastly different from Earth. The 0.9 tonne Curiosity Rover (Mars Science Laboratory, MSL[84]) is the largest payload we've ever landed on Mars. The problem is that because of the thin Martian atmosphere, we can't use the same technologies to land larger payloads as we've used for robotic missions. The fact that Mars has an atmosphere is very nice from the perspective that we can use atmospheric drag to slow approaching spacecraft. However, this drag doesn't slow spacecraft enough, and terminal velocity is therefore higher than on Earth[85]. Additionally, the thin atmosphere means that parachutes are less effective.

One potential solution to this problem is to rely more on retro rockets. This is the approach currently being investigated by SpaceX for the Dragon landing system. "Red Dragon", planned for 2017, would deliver a capsule filled with scientific instruments to drill underground and look for reservoirs of liquid water. Drawbacks of using retro-rockets are the

---

[83] The speed of sound on Mars is around 878 km/hr (536 mph), as compared with 1236 km/hr (768 mph) on Earth. Due to the thin atmosphere, if you could talk on Mars, your voice wouldn't carry far and it would sound high pitched like a Smurf, or as if you had inhaled a whiff of Helium. On Venus, your voice would carry further and sound deeper, as if you had inhaled Argon. Incidentally, Mars probably smells like iron rust.
[84] Another TLA!
[85] Woe to the casual Martian skydiver, who would reach a terminal velocity of around 928 km/hr (580 mph), as compared with a terminal velocity on Earth of 195 km/hr (120 mph). Of course, terminal velocity depends on mass and shape, so varies by falling object. For example, a feather has a very slow terminal velocity.

fuel required, and potential instability from firing rockets in hypersonic conditions. However, this problem may not be insurmountable: SpaceX is planning to test supersonic retro-propulsion later this year, which could be used on both Earth and Mars.

Another alternative is to use a hypersonic inflatable system instead of a parachute. These are basically gigantic balloon-like objects that would expand and stiffen to become something like a super-rigid upside-down parachute that decelerates while also providing a heat shield. NASA has tested these a couple times, though has never used them on an actual mission. Built of high strength, lightweight, and flexible materials, this kind of technology could be ideal for a Mars mission.

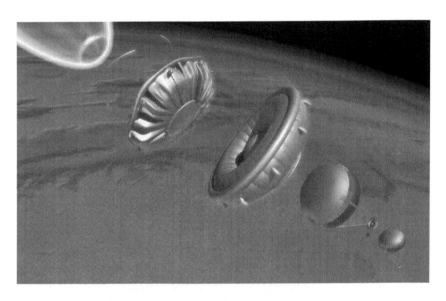

NASA's hypersonic inflatable deceleration system

Mars One

So what is the actual Mars One plan? The current plan seems to rely on being able to purchase Falcon Heavy launches and send the crews in upscaled Dragon capsules. As such, the spacecraft design, manufacture, and operation would rely almost completely on SpaceX. Other companies would provide life support systems and inflatable modules. According to the Mars One roadmap (Fig. 3), a demonstration supply mission would be flown in 2016, presumably with a Falcon Heavy and Dragon. This mission would carry supplies, both as a concept demonstrator, and because if you're sending something there anyway, why not make use of it? In 2018, a support rover would be launched to identify a suitable site for a base near accessible water[86]. In 2020, the two redundant sets of base components would be launched, along with two rovers and backup supplies. These would be spread across six launches in total: two habitats, two life support units, and two supply missions. These base units would be assembled by rover on Mars.

If anything goes wrong with a prior mission, you delay sending humans. Any crews leaving Earth head for Mars with the full knowledge that there is a (relatively) safe shelter waiting for them, along with at least two sets of spare parts, and backup supplies to last for years. Every two years, starting in 2022, a crew of 4 would head for Mars.

---

[86] It strikes me as odd to launch the supply mission before the rover, because the rover's task is nominally to identify a good location to place a base - where you would want your supplies. However, since the most important part of the mission is being able to successfully and reliably land large payloads on Mars, perhaps it's best to demonstrate that capability fist even if the supplies end up far from the base. No large payloads on Mars = no humans on Mars.

Bearing in mind that everything always seems much easier on paper, it is a fairly robust – though extraordinarily ambitious - concept. The acute dangers stem from launch failure on takeoff, mechanical failure during space travel, or an accident while trying to land on Mars. However, these dangers will be present for any Mars mission, and the required capabilities will have been demonstrated multiple times with similar hardware. Considering this, is it really more dangerous than the Moon missions were? The Moon landing capability was remotely demonstrated precisely zero times in advance, and only partly demonstrated with humans once, on Apollo 10. When Apollo 8 carried humans to the Moon, it had never before left Earth orbit[87].

Once the base modules are landed, solar panels would be deployed for power[88], and the base would start stockpiling water, oxygen, nitrogen, methane, and hydrogen. Since Martian soil[89] is thought to be around 5% water ice by weight, extracting water on Mars is relatively

---

[87] People often forget just how little testing was done before we sent humans to the Moon. Saturn V flew only twice before it carried people to space! The Apollo command spacecraft was tested 5 times, and the lunar module was tested in Earth orbit twice: once unmanned, and once manned. It's amazing what a deadline can do.

[88] I have not seen the numbers, but I suspect that it might be difficult to rely on solar panels alone. However, there are serious political challenges involved in launching anything nuclear into space, considering fears of radioactive release into the atmosphere. These fears seem mostly groundless to me. Although it is true that just small amounts of plutonium would be toxic enough to kill everyone on Earth, the distribution problem means that this simply wouldn't happen. Additionally, all the spacecraft we've launched to the outer solar system already carry plutonium in their Radioisotope Thermal Generator (RTG) power supplies. But don't tell anyone.

[89] What shall we call dirt on Mars? The technical term for "soil" on another world is "regolith", which is what I tend to call it, but I don't want to confuse anyone. "Dirt" seems a little abrupt. Obviously we can't call it "Earth", and calling it "Mars" would be really confusing. So I guess I'll call it soil, even though that usually means dirt with at least some biomass, which is probably not the case here.

straightforward: scoop up some regolith, melt it in a heater, and filter out the water.

Assuming all of this goes according to plan, the first crew of 4 would launch in 2022, arriving at Mars in 2023. It takes between six and eight months to get to Mars. This is certainly a long flight, but not substantially longer than a tour on the International Space Station, and significantly less than the longest stay on the Russian space station Mir[90]. According to the Mars One website, the transit habitation will "feature" around 20 m^3 of living space. This is small, but not ridiculously so[91].

Even so, the long space voyage will certainly not be a luxury cruise, but humans have experienced much worse. The explorers that sailed with Columbus across the Atlantic in 1492 had less than 10 m^3 per person, and submarine crews before the nuclear age had even less[92]. The assembled Mars transit vehicle would consist of a transit habitation, two propellant stages, and a landing capsule. With backup supplies and a crew of 4, this would significantly exceed the volume and mass capacity of a Dragon/Falcon Heavy combination. Therefore, multiple launches with rendezvous in Earth orbit would be required for this kind of architecture.

When the crew lands on Mars, they would find a fully assembled and checked out base with backups for all systems. Should the crew fail to land near the base, their rover would be able to travel several kilometers.

---

[90] The longest single flight in space was by Cosmonaut Valeri Polyakov, who spent 437 days in space (over 14 months) on Mir in 1994-1995.

[91] By comparison, a typical minivan has a total interior volume of 5-7 m^3 or so – so each person gets 3-4 minivans worth of room - certainly not crazy for a road trip this long. However, this is about one-tenth of the pressurized volume of ISS, at almost 900 m^3.

[92] If you think this is bad, Jim Lovell, the commander of Apollo 13, spent 14 days cramped together with Frank Borman in a capsule with around 2 m^3 of space total during Gemini VII – this is cramped for the front seat of a car. No showering, no changing clothes, and practically no stretching of any kind. Not for the claustrophobic, or those sensitive to odours!

To account for more extreme cases, each crew launch would be paired with a supply ship. This would follow the crew by several days, so it could be redirected to the crew landing site if necessary. Although we have lost many robotic Mars missions during landing[93], our historical track record of being able to land at designated points on Mars and the Moon has been excellent – and these are the types of dangers that any Mars mission will face.

The plan is rather robust for a mission of this scale. New technologies will not dramatically reduce these risks. Moreover, the plan becomes more redundant over time. Each time you send a crew, you also send extra base components and backup equipment. Eventually you have a small but growing settlement on Mars.

---

[93] Following the Soviet/Russian proclivity for more and cheaper but less reliable missions, the USSR/Russia has launched 21 Mars mission, of which only 2 successful and 2 partially successful, or a success rate of between 10-20%. They were more successful with Venus, with 16 successes and 15 failures. I guess you've got to admire their tenacity. The US track record is almost the reverse, with only 5 failures and 17 successes. I just hope the Russians aren't doing the Mars trajectory planning...

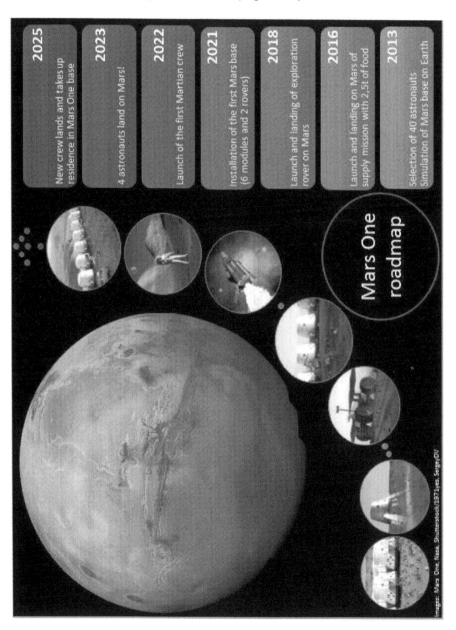

The Mars One roadmap, as illustrated on their website

Living on Mars

Ok, we're on Mars surrounded by seemingly endless rusty-brown desert. How the heck do we live here? Living on Mars isn't all that different from living in the spacecraft, except hopefully there would be more interior volume. According to the Mars One website, "the Mars habitat will be a modular environment made up of multiple inflatable units, and will comprise about 1000 m^3 of total living space, which equates to 250 m^3 per inhabitant for a team of four". This is a decent sized house, and more volume than the International Space Station. So you'd have plenty of room for your own space, in addition to common areas, and specialized laboratories. The habitat would be modular so that even if one inflatable unit is damaged beyond repair, the habitat would still be functional, and the crew could make do until a replacement was sent.

The whole point is that you go out and explore. Due to the low pressure, you'd have to wear a space suit whenever you went outside[94]. You would be able to communicate with Earth using a relay satellite, just like Curiosity rover does now[95]. There is a time delay so direct phone calls are impractical[96], but email, photos, and recorded video would be fine. This is much better service than the early settlers of North America had.

---

[94] The space community calls going outside "extravehicular activity" or "EVA". Another TLA, and I'll try not to use it, but no promises.

[95] Mars One has stated that they plan to launch their own communications relay satellite. For redundancy and coverage, you might need at least two.

[96] And extremely annoying – you'd have to wait up to 48 minutes for a response! Also, with the time delay, Internet experience would be highly compromised, meaning that most young people probably wouldn't want to go – no smartphones on Mars! Although perhaps relevant webpages could be cached and the Internet could be accessed in a limited fashion. Kind of like the 1990s.

Martian bases would follow a progression over time. Early habitats would be clustered around the arriving space capsules, supplemented by inflatable structures brought from Earth. Martian soil would cover the base for radiation shielding, and could also be made into bricks. Later habitats could be built from these bricks, and could be dug into the ground or the side of a hill, possibly even taking advantage of natural features such as caves[97].

Plastics manufacturing on Mars is straightforward, and this technology could be used to expand the base by appending extruded or inflatable modules. Eventually, the construction of pressurized domes would become feasible, and in the very long run we can imagine pressurized domed towns and eventually cities on the surface.

An early base might consist of connected space capsules and inflatable structures

---

[97] This is pretty hard work, especially in a space suit. However, perhaps a rover would be designed to do this.

Water and carbon dioxide would be converted to provide breathable air and fuel. Glasses, ceramics, and natural fibers like bamboo and hemp are also near-term possibilities. Although the base would be dependent on Martian water and air production, most technological products like computers would be sent from Earth. Some critical components could be produced or replaced: the base would certainly have 3D printers[98].

The Mars One plan calls for food production on the first mission. While this should be attempted, I am unconvinced that it would be possible to grow much beyond a supplement in the first year or two. As such, early supply missions would likely have to carry adequate food for the entire duration, along with backup emergency rations. While certainly useful for vitamins and morale, even the food varieties that could be grown early-on would be limited, so a Mars base would still probably be dependent on at least some dried food from Earth for a while[99].

However, in the longer term, food production is going to be essential to live on Mars sustainably. An inflatable greenhouse could pump in Martian carbon dioxide, although insulation would be required. Hydroponics and aeroponics[100] would be especially useful. You'd want to focus on easy plants like the ones you grow in your garden: lettuce, tomatoes, peas, beans, carrots, radishes, strawberries, and cabbage.

You would certainly not bring any animals on early flights, and it is probably better in the long run if meat consumption were heavily

---

[98] At least two 3D printers: for redundancy, and for when Fred is printing up his own action figures. This technology is scheduled for testing on ISS starting in 2014.

[99] I hope they send lots of astronaut ice cream. It's my favorite food.

[100] Hydroponics = growing plants in water. Aeroponics = growing plants in air using a mist. The US military conducted extensive experiments on hydroponics and aeroponics with great success to support their far-flung bases around the world during the Cold War. This technology would also be applicable in urban environments on Earth, for example, in vertical farming using old office buildings to grow many floors of plants.

restricted[101]. Even if you did bring animals on a later flight, the best kinds would be small and efficient at converting food into meat[102]. Protein might come from sources such as spirulina, a blue-green algae and probably the highest protein food you can eat[103].

Mushrooms might also be grown – these provide some of the same B vitamins that people get from eating meat. Although it sounds offensive to North American sensibilities, another great dietary option would be insects. Over a thousand species of insect are consumed by billions of people, in 80% of the world's nations[104]. If water wasn't an issue, aquaculture might be used to harvest fish and shellfish.

We have the technology to keep people alive on Mars. There are surely many unforeseen challenges and unanswered questions, but confronting them is the best way to solve them. We have a lot to learn, so let's get going.

---

[101] That's ok for me, I'm mostly vegetarian anyway – it's called "flexitarian".

[102] In the early space program (a much more meat & potato era), they studied what types of animals would be the best to bring, in terms of minimizing waste, effort, and feed. It turns out that mice made the best candidates. Mmmm... mouse stew, just like Mom used to make. For more fun space facts like this, check out Mary Roach's excellent book *Packing for Mars*.

[103] To be honest, it kinda tastes gross, and large amounts are unpalatable. I tried baking spirulina into "Moon cookies", which was only partly successful. But you're not going to Mars for the food. You're going for the sunsets.

[104] Come on, where's your adventurous spirit? I've eaten insects a number of times. They don't taste great, but they're food. Anatomically, they aren't very different from shrimp or lobster, so what are you afraid of?

Unresolved Issues

There are several interesting long term questions regarding the settlement of Mars. Would the settlers be able to have children? In the early days, a Mars base would not provide a good environment for child rearing, so this would probably be discouraged, but in the long run, the survival of the Martian community would of course depend on new generations being born there. Is it even possible to have children on Mars? Strictly speaking, we don't know. Experiments involving rats on the Russian Biosatellite, and on amphibian sperm and egg cells have been inconclusive, but suggest that fertilization is possible in zero gravity[105].

There may be issues with egg implantation in the uterus, and as for fetus development, we really don't know. Since a fetus floats with neutral buoyancy in fluid anyway, there may not be a huge impact. On the other hand, it is possible that there would be unforeseen health impacts, and certainly children developing in zero gravity would be weak to the point that they might have trouble ever returning to Earth.

However, babies go through various stages of locomotion, and their muscles don't spontaneously develop – rather they develop in proportion to their activity. Before babies crawl, their arm and leg muscles are very weak. Until they walk, their legs remain underdeveloped. Thus, perhaps returning Martians could at least partially readapt to Earth. Moreover, we're not talking about zero gravity, we're talking about Mars. It seems obvious that some of the reproductive and development challenges

---

[105] Technically, nowhere is "zero gravity" – it's just that the acceleration of an orbit cancels the gravitational force. But I'll say zero gravity for simplicity's sake.

associated with zero gravity would apply less, if at all, to a Martian environment with 38% gravity.

There are potential legal issues as well. Launching a space mission from the United States requires a launch permit from the Federal Aviation Administration (FAA). Even if the actual launch were planned on a spacecraft rated to carry humans, is there some chance that the US government would attempt to intervene? Another potential hurdle is that the United States restricts the foreign sale of space and military hardware under the terms of 'International Traffic in Arms Regulations' (ITAR). What impact would these have[106]? Who is responsible for the support and well-being of the Martian settlers[107]? Once the settlers got to Mars, what laws would apply? Would they even have the legal authority to establish a base on another planet?

The space treaties signed by many nations, including the United States, prohibit private ownership in space. This means that there might be legal troubles with claiming land, or for example, selling Mars rocks on Earth. However, Mars One is not alone in facing these challenges. Newly emerging asteroid mining companies like Planetary Resources may have to deal with these issues first, and the resolution would certainly have repercussions for Mars[108].

Since one of the main reasons to go to Mars is to search for life, another potential issue with having humans on Mars is contamination. Although some bacteria have been shown to be able to survive in space for long

---

[106] Perhaps none if the mission were entirely run from the United States and the technical aspects were managed by a US prime contractor like SpaceX.
[107] My view is no one, really – they understand the risks and take their chances. But this might not fly legally. And it is an interesting question of whether the "stranding" laws would apply. Legally, countries are responsible for rescuing people in distress in their territory, for example, sailors lost at sea.
[108] I suspect that these treaties were easy to sign because they had no actual near-term impact. Once these issues start to come up in practical terms, they will surely be re-examined.

periods of time, it is far easier to decontaminate a few robotic explorers than to prevent human contamination of a Martian environment. Nevertheless, the risks of human contamination are often overstated. Most Earth microbes would have great difficulty surviving and reproducing on Mars, and the chances that they would be spread planet-wide are extraordinarily slim[109].

Moreover, the most likely candidate areas for life, such as underground reservoirs, would almost certainly be insulated from contamination unless they were very close to the human settlement. Even after humans are well established, the 2011 United Nations Committee on Space Research (COSPAR) panel suggested that it would be sufficient to designate certain areas of Mars "protected zones" where humans would not be allowed to venture[110].

While we are on the subject, there are several other interesting questions about Martian life. First, could it pose a threat to humans? This would be extraordinarily unlikely. Bacteria and viruses on Earth only infect humans if they have specifically evolved to do so – no humans, no human diseases. X-Files aside, this is the reason we would never find a human disease at the bottom of a mine shaft or volcanic vent.

Second, if we found Martian life, would we have a responsibility to preserve it? Although I would give a definite 'yes' on the basic issue of that life as a scientific treasure, I would vehemently say 'no' on any issues of morality. If there were intelligent or even multicellular life the answer would perhaps be different, but we are almost certainly talking about nothing but microbes.

Third, where did the life come from? The main question we would want to answer is if Martian life was completely independent from Earth life.

---

[109] In fact, if they survive and reproduce, this might be a good thing because it would demonstrate that life can indeed survive on Mars.
[110] Indeed, the Chair of COSPAR is an advisor to Mars One.

We know that small meteoroids are occasionally transferred between the two planets as a result of impacts blasting pieces into space, and that microbes can survive in space.

So did Earth life originate on Mars, or vice-versa? Or did the two planets have completely different origins of life? This case would be even more interesting, because it would suggest that the spontaneous instantiation of life is not as rare as it might seem. Perhaps life is common throughout the Universe.

Paying for Mars

No matter how you do it, sending humans to Mars will cost a lot of money. Even though going one-way is certainly far less expensive than a traditional approach, it will still cost billions of dollars. Cost estimates for a NASA Mars mission have typically ranged somewhere in the neighbourhood of several hundred billion dollars, spent over a decade or more[111]. Using private spacecraft and components may reduce the cost from hundreds of billions of dollars to tens of billions, but funding is always going to be the biggest challenge facing a Mars mission.

Mars One estimates that the first mission would cost on the order of $6 billion, with $4 billion for follow-up flights. Considering the hardware they plan to use[112], this price is not outside the realm of possibility[113]. Since some of the hardware hasn't even been designed yet, and space projects always cost a lot more than expected, this estimate is almost certainly on the low side. Still, it is conceivable that such a mission might be possible for an initial cost of less than $20 billion, and perhaps even closer to $10 billion. How would it be possible to raise this much money?

---

[111] How much money is this? To give some idea, it is around 1/60th of yearly US GDP, 1/12th of the yearly federal budget, 1/3rd of the cost of the Iraq War, 1/2 of the yearly US defense budget, or ten times NASA's yearly budget. But this cost would be spread over a decade or more, meaning that the cost of even a NASA mission to Mars is not extreme – just too high to make it politically appetizing.

[112] Around 12 Falcon Heavy/Dragon launches for the first mission, at something like $300 million each.

[113] Inspiration Mars, the planned 2018 Mars flyby, projects a cost of around $2 billion using similar hardware.

Though a very large sum, it may not be beyond our appetite for expenditure. It represents only a couple years of what NASA spends on human spaceflight[114], and less than a tenth of the total cost of the International Space Station[115]. Looking at other examples, Boeing 787 development cost around $32 billion, with each plane costing around $250 million – not much less than the projected cost of a launch to Mars. The newly planned Istanbul airport is projected to cost $30 billion, about the same price as China's Three Gorges Dam.

Other notable megaprojects in this price range include (2013 dollars): the Manhattan Project ($26 billion), the English Channel tunnel ($17 billion), the Boston "Big Dig" tunnel system ($15 billion), the Arabian Canal ($11 billion), the John F. Kennedy airport expansion ($10 billion), the Yucca Mountain nuclear waste depository ($9 billion), and the Atlanta-Jackson airport expansion ($9 billion).

Compared with military price tags, a Mars mission could be a bargain: the Gerald Ford class of aircraft carriers under construction are projected to cost over $9 billion each, in addition to over $14 billion in development. They have a daily operating cost of $7 million – adding up to over $50 billion for 10-years of operation for the pair.

Mars One seeks to fund the mission to Mars as the "largest media spectacle in history". It would certainly have to be. The example they cite is the Olympics, which earns over a billion dollars per week from media rights. With a total cost of around $14 billion, London was close to breaking even. It is interesting to reflect that if Mars One is anywhere close in their cost estimates, London could have decided to launch a Mars mission instead of organizing the Olympics.

---

[114] Around $6 billion per year.
[115] Over $150 billion.

Like the Olympics, but unlike any of the aforementioned megaprojects, Mars One would be able to leverage financial support from entire world. It is not a US, nor European, nor Russian, nor Chinese project. A landing on Mars will be a pivotal moment in human history, and like the Moon landings, it is something the entire world would watch. I think it's safe to say that global public interest for an actual mission to Mars would be higher than the Olympics, so earning enough money to pay for the mission might be possible. Even so, the cash flow timeline doesn't really work. Assuming all went according to plan, revenue would pour in years after most of the hardware purchases. Thus while selling Mars media rights is a good plan, it is far from a complete plan. Generous and optimistic investors, both big and small, will be necessary to bridge this gap.

Remember of course that Mars One plans not only to sell media rights to the actual mission, but to the crew selection, preparation, and training before the mission, starting around 2015[116]. This is actually the critical part. The top rating television show in the US for many years was American Idol, which earned around a billion dollars (2012 dollars) per year. American Idol was broadcast mainly in the United States, but Mars One would operate on a global stage, having access to a market several times as large.

A mission to Mars would provide absolutely unique content that could be spun into a variety of formats, both real and parody. Thus, it is hard to see why there would be any major obstacles in funding a television show that sincerely plans to land humans on Mars. It would not pay for Mars, but it might just attract investors. The success of the show is absolutely critical. If the TV show flopped, it would likely scare away any would-be investors. If, on the other hand, the precursor TV show proved an

---

[116] There is a very interesting question about how a TV show would be presented: it may prove difficult to balance entertainment and technical content.

international smash hit, this might generate revenue and attract enough investors to proceed with subsequent phases of the mission[117]. I wonder if it's a sad commentary on our society that a reality television show could have so much leverage on the future trajectory of our civilization.

---

[117] There would obviously be a lot of pressure to keep content entertaining. However, don't expect Jersey Shore on Mars – my guess is more like National Geographic with attitude. No Snooki here, but perhaps a Steve Irwin.

Attracting Investors

Mars One is a registered non-profit which seeks donations from around the world. So far they have received over $100,000, in addition to revenues from merchandise sales[118]. Since application fees were an average of around $35, with over a hundred thousand applicants as of August 2013, Mars One has already pulled in millions of dollars. Although this is a far cry from the billions required, there does seem to be a lot of public interest and media attention. It may be possible to raise quite a sum from small-scale funding.

On a very simplistic level, this model could work. If everyone in the world donated just $1 to the project, this would be $7 billion dollars[119]. This might not completely pay for the mission, but it could certainly bridge the funding gap. One or more online crowdsourced campaigns might also be an option.

The most successful Kickstarter campaign to date raised over $10 million for the Pebble smartwatch, and there have been about a dozen campaigns in the $3 million dollar range. A recent Planetary Resources campaign raised over $1.5 million for the ARKYD space telescope, indicating that space crowdsourcing projects can work. Even though none of these projects would provide more than a small fraction of the funds required for Mars One, crowdsourcing could be a slice of the funding pie. And who knows? Maybe a mission to Mars could just turn into the most successful crowdsourced project of all time.

---

[118] I currently donate a dollar per day, in addition to $1 per copy of this book sold. Tell your friends – if everyone buys this book, we're going to Mars!
[119] Or more realistically, if the 10% or so people of the people in the world who can afford to donate and believe in a mission to Mars contributed just $10.

Alternatively, there are over 850 billionaires in the world. If just a few of them decided to back the project, it could absolutely work. There are over 90 people in the world who could pay for the entire mission themselves, even assuming the costs estimates were doubled. Bill Gates, or the richest man in the world, Mexican Carlos Slim, could afford to fund several missions like this. It is not impossible that a billionaire would decide to fund the first human mission to Mars, either to secure their legacy, or simply to change the world[120].

It is also possible that potential media partners seeking to profit from the mission in the long run might be willing to fund portions of the mission in advance. Google seems like an obvious candidate, with both the financial resources, and co-founders Sergey Brin and Larry Page who are actually interested in getting humans to Mars. Apple, Microsoft, or more traditional media companies like the big six (Comcast/NBC, Fox, Disney, CBS, Time Warner, and Sony) would also be candidates. However, neither billionaires, nor giant media companies got to where they are by making frivolous and unnecessarily risky investments. Thus, this kind of large investment may not be forthcoming until after feasibility becomes clear, particularly after the 2016 demonstration mission.

It is also likely that commercial advertisers would be interested in the mission. Although the risk may dissuade some, the physical danger might actually attract certain brands. Red Bull recently funded the record parachute jump, carried out by Felix Baumgartner from over 110,000 ft, and Unilever (AXE in North America, or Lynx in Europe) has purchased 22 suborbital flights as part of a marketing campaign. How much would Coca-Cola pay to brand a mission to Mars?

---

[120] Inspiration Mars, the 2018 flyby is backed by Dennis Tito, and though his fortune is not enough to pay for the estimated $2 billion, it is sufficient to pay for early stages and fund an effort to raise additional funds.

Sustaining Mars

Assuming Mars One was able to raise enough capital to fund the first mission, would they be able to fund follow-up missions? Would a humans-to-Mars mission be able to retain the attention of the world, or would the public lose interest, as eventually happened with the Moon landings? Although it's hard to imagine a scenario where the majority of the world wouldn't watch at least the initial footfalls, it will be a challenge to keep people interested. However, there are several advantages as compared with the Apollo missions. As a private mission, there would be fewer constraints on how content was presented.

Moreover, this is a mission of settlement, not a short "flags & footprints" turn around mission. The real human drama only begins once the colonists land on Mars. How will the settlers establish their base, harvest their air and water, grow their food, and survive? How will humans fare exploring a planet with as much land as the entire Earth? Psychologically, how will humans react to continual confinement together? What will happen when the follow-up crews, still training on Earth, finally arrive? What will happen when the first children are (eventually) born on Mars? What interpersonal dramas will play out? This is a human story, one likely to maintain interest.

Even if attention wanes, a significant proportion of the global population supports sending humans to space. While not able to repeat Apollo, NASA has managed to maintain steady funding levels since. Settlers on Mars will be able to communicate with the entire home planet, and new social media technologies like Twitter make this even easier. Chris Hadfield was able to amass over a million followers, and that was just from Earth orbit. Imagine how these kinds of platforms could reach out and inspire our youth, with classroom videos and question & answer sessions.

Follow-up flights will be cheaper and easier, and eventually the base on Mars will become more self-sufficient[121]. Until then, crews on Mars would be vulnerable to funding shortfalls. In such cases, it might not be possible to fund follow-up missions, but it is highly unlikely that sufficient funds would not be available to pay for resupply flights.

Say we have 20 people on Mars after 10 years. A resupply flight would cost around $250 million. Let's double that for redundancy (2 flights) to $500 million, every two years. On a global scale, or for a large corporation, this is nothing. This is 10% of what NASA currently spends on human spaceflight per year.

Since Mars One would be a highly international mission, with settlers drawn from many nations, even in dire straits each country would only have to pitch in $12.5 million to sustain the Martian settlement. This is less than many nations (and indeed wealthy individuals) are currently willing to spend on a flight to the International Space Station. Even if public interest declined significantly, it is hard to imagine the media business case entirely evaporating.

Moreover, there are other potential revenue streams. With humans on Mars, experiment time could be outsourced to science laboratories on Earth. Telerobotic rover operations would be extremely easy, ranging over the entire planet. Planetary scientists could buy time on the Mars One communications satellites. Samples of Mars could be easily sent back to Earth, both for scientific study, and also potentially to sell to collectors[122]. Considering that a pair of Moon rocks recently fetched over

---

[121] Though not entirely self-sufficient for many years.
[122] Although whether there are legal issues with this remains to be seen.

$330,000 at auction, Martian rocks could be worth quite a lot[123]. How much would you pay for a piece of Mars?

A piece of Mars on Earth: the ALH84001 meteorite recovered from Antarctica in 1984. In 1996, scientists announced that it showed evidence of fossilised life forms. However, there is no scientific consensus.

---

[123] There is also some evidence that Mars may possess abundant diamonds, based on high concentrations of carbon in the mantle, and volcanic activity that would bring them to the surface. But you can't count on something like this.

Risk

I am not by any means a risk taker. Probably the riskiest thing I have ever done[124] is skydive, and this was only because it was part of a challenge on the reality TV show where I appeared in early 2013[125]. Although I have a pilot's license and am an aviation enthusiast, I'd never given a lot of thought to skydiving.

Hesitant at first, when put to the test, and assigned a mission to remember a dozen letters spread out on a landing strip, I decided to go for it. Admittedly, part of it was peer pressure – I was certainly conscious that I was on TV. Others on my team were also nervous, some probably even more than I. Seeing my ex-Air Force teammate Beth jump first helped a lot (I wonder what the first skydiver was thinking?[126]). But mostly, I decided that it was worth the risk in order to accomplish something, even if that something was no more than winning a challenge in a competition.

Having a mission to accomplish, and a task to focus on, goes a long way towards justifying risk for me. In fact, the main thing that was going through my mind at the time was "If I die here in a stupid skydiving accident, I'm never going to make it to Mars...". I even remember discussing the relative risks and potential benefits of a trip to Mars vs. skydiving with the television host.

Clearly, going to space is much, much riskier than skydiving. Just how risky is it? It is currently one of the riskiest jobs you can do, at least in

---

[124] Apart from living, which still unfortunately carries a mortality rate of 100%

[125] Canada's Greatest Know-it-All.

[126] Actually, the first parachutist, Franz Reichelt, was killed on his first jump off the Eiffel tower in 1912 -- I wonder what the second thought!?

the western world. As a species, we've sent over 500 people to space, and at least 18 have been killed there[127], with at least 10 more being killed in training[128]. This means that as an astronaut, you have something like a 5% chance being killed on the job. How does this compare with other risky careers?

As of 2013, over 2,000 US soldiers have been killed in Afghanistan, with a peak deployment of over 100,000 troops. If you don't consider rotations, this means that over a 12-year time frame (a typical astronaut career), the equivalent rate is roughly 2%. The risk per soldier is lower if you account for the fact that troops are rotated, however, military training accidents are not uncommon, and soldiers often see multiple tours or action in other theaters such as Iraq (with a similar deployment loss ratio of 2.5%)[129], which somewhat offsets this.

Over a 21-year period from 1990 to 2011, encompassing both relatively quiet and active times, the average death rate for US active duty military personnel from all causes was 72 per 100,000 person-years[130]. This means that over a 40-year career, your chance of being killed in service was about 2.8%. However, this is an average across all branches and duties. Clearly, it would be far less risky to perform the majority of support roles, and far more risky to be in front-line infantry service. Additionally, this figure fails to account for that fact that the chance of

---

[127] They were: Vladimir Komarov (1967), Georgi Dobrovolski, Viktor Patsayev, & Vladislav Volkov (1971); Greg Jarvis, Christa McAuliffe, Ronald McNair, Ellison Onizuka, Judith Resnik, Michael J. Smith, & Dick Scobee (Challenger, 1986); Rick Husband, William McCool, Michael P. Anderson, David M. Brown, Kalpana Chawla, Laurel Clark, & Ilan Ramon (Columbia, 2003).

[128] They were: Valentin Bondarenekoi (1961); Theodore Freeman (1964); Elliot See & Charles Bassett (1966); Virgil Grissom, Edward White, & Roger Chaffee (Apollo 1, 1967); Clifton Williams (1967); Robert Lawrence (1967); Yuri Gagarin – first human in space (1968); Sergei Vozovikov (1993).

[129] US losses in Iraq have been over 4,500 killed with over 32,000 wounded for a peak deployment over 175,000.

[130] Medical Surveillance Monthly Report, Vol. 19 No, 5. May 2012.

being wounded is seven times as high as that of being killed[131], meaning that compared with an astronaut, even if your chance of being killed is lower, your chance of being wounded is much, much higher in the military.

In the civilian sector, things aren't actually better. According the Bureau of Labour Statistics (2011), and assuming a career length of 40 years, the ten riskiest jobs by occupational death rate per worker over a career are: fishermen (4.6%), logging workers (3.6%), aircraft pilots (2.8%), farmers (1.6%), roofers (1.1%), miners (1.0%), waste collectors (1.0%), truck drivers (0.9%), machine operators (0.8%), and police & firemen (both around 0.7%). Kind of makes you think twice about becoming a fisherman, doesn't it?

We should also keep in mind that although these are currently amongst the riskiest jobs in the western world, they are nowhere near the riskiest jobs of all time, nor even the riskiest jobs in the world today. These figures are for the United States, where figures are reliable and conditions are relatively good, even for risky jobs. The equivalent rates for civilian jobs in developing countries around the world must surely be higher - and just think of how much more dangerous it would be to be a soldier fighting in a third world conflict zone, or even be a civilian living there.

Obviously, we don't want to voluntarily subject people to more risk than we have to, but it is still illustrative to reflect upon some examples from history. Over 3 million soldiers fought during the four years of the US Civil War, and over 600,000 died in it, along with another 450,000 wounded. This is a relative death rate of around 20%, for just four years of fighting. Yes, life was shorter and harder back then, but this astonishing rate - more than ten times what we see today – just goes to

---

[131] Coalition losses in Afghanistan to 2013 have been approximately 3,313 killed plus more than 23,500 wounded.

show you the level of risk we have tolerated in the past when we thought it was justified. For a 20th century example, the US lost around 5% of men deployed in the Second World War[132], and these figures would have been much higher had the scheduled invasion of Japan taken place[133].

Moreover, US losses were extremely mild compared with most. The Soviet Union lost as many as 36% of deployed personnel – almost 9 million men, or roughly the population of Sweden[134]. Poland has the dubious distinction of having lost the largest population percentage, with around 16% killed, counting both civilian and military deaths[135]. Indeed, during the Second World War, more than 2.5% of the population of the entire Earth was killed!

So what's my point? Obviously, no one wants to recreate the conditions of the US Civil War, the Second World War, or any other conflict. My point is that in times of crisis, humans have tolerated extreme sacrifices and levels of risk to accomplish major goals, such as maintaining the Union and ending slavery, or defeating the Nazis and Imperial Japan. We may not think about it in our comfortable lives, but if you were shopping for a job throughout human history, you could do a heck of a lot worse than going to Mars, even one-way.

The history of exploration isn't much better. Columbus' ships were tiny craft less than 60 ft long, and they sailed into the unknown without any of the modern risk assessment we would expect today. His largest ship, *Santa Maria*, was wrecked along the way, and he had to leave men behind in the New World. Magellan's expedition, sailing around the

---

[132] Around 420,000 were killed from around 8.5 million deployed.
[133] Projected US casualties were in the hundreds of thousands. Of the 500,000 Purple Hearts manufactured in expectation of the 1945 invasion, over 100,000 have yet to be given out as of 2013 – they are so overstocked that units carry around spares for immediate award in the field.
[134] Of around 25 million deployed, in addition to at least 15 million more civilian deaths.
[135] Around 5.5 million of a population of around 35 million.

world for the first time starting in 1519 was perhaps the greatest exploratory achievement of all time. However, of five ships that set out with 237 men, only one ship, *Victoria*, completed the circumnavigation and struggled back to Europe with only 18 men on board[136].

Route of the *Victoria*

Franklin's 1845 expedition to discover the Northwest Passage was lost with all 134 souls when the ships, *Erebus* and *Terror*, froze into the ice[137]. Ernest Shackleton's ship *Endurance* became similarly stuck in Antarctic ice during his 1914-1917 expedition. However, thanks to his

---

[136] One ship, *San Antonio*, had previously abandoned the expedition and returned to Spain. Magellan himself was killed by natives in the Philippines when he meddled in local politics.
[137] This was during the early days of using canning for food preservation, and it seems that lead poising and the associated symptoms - including insanity - may have played a role in the fate of the expedition.

leadership, he managed to save all his men after what must have been one of the worst ordeals in exploratory history[138].

Looking back on these exploratory endeavors with a modern eye, they no doubt seem reckless – but would we consider them to have been mistakes? Would we prefer that they had stayed at home where it was safe?[139]

I certainly would not: these explorers had the bravery to do what they could to change the world, even if they were often motivated by dreams of personal advancement. If explorers of centuries past had insisted on a 99% or even 90% chance of return, the whole course of human history would have been dramatically different and we would not be where we are today.

I am by no means suggesting that we assume risks as recklessly as those we accepted throughout the history of exploration. I am simply suggesting that, compared with the risks we have already taken, a human mission to Mars would not rank at the top of the list, either in terms of risk to the individual, or in terms of total lives put at risk. As perhaps the explorers of the past understood far better than we do today, great achievement is worthy of great risk.

In comparison with what our ancestors have done, a one-way mission to Mars sounds downright reasonable. No flights would be sent to Mars before multiple successful demonstrations of the same launch and landing technologies. No crew would head off to Mars until there was a

---

[138] Shackleton's expedition earns the title of "most successful failed expedition". Like Apollo 13 it was meaningless in exploratory terms – but as a human story of courage, teamwork, and endurance, it is one of the greatest of all time. Check out *Endurance: Shackleton's Incredible Voyage* by Alfred Lansing.

[139] Not so safe for some of course, considering Europe during the age of exploration was a cauldron of war and disease, with a life expectancy of around 35.

base already in place, with system checks complete and everything backed-up with spare parts. How Bering and his Russian explorers would have envied their Martian counterparts.

There are certainly risks, but we simply can't apply same standards of risk to exploration as we do for routine operations[140]. While a 5% chance of fatality is far too high for a mission to the International Space Station, it should be perfectly acceptable for a mission to Mars. These days, we seem incapable of making this distinction.

So what are the major risks of a Mars mission? There are several types. There are of course risks associated with all spaceflight, particularly with launch and landing. Then there are the risks associated with operating complex equipment far from Earth, for example micrometeoroid impact[141], or multi-system mechanical failure. Finally, there are the risks associated with having humans in space and on Mars for long periods of time, both physiological[142] and psychological.

---

[140] Some would argue that any human space operation can be called "routine", but this is precisely my point. For space exploration, fewer, riskier, and more ambitious missions are preferable to a large number of safer mission with less ambitious goals.

[141] This is actually rarer than most people think. Space is big and mostly empty. Compared with interplanetary space, Earth orbit is utterly filled with debris, and significant collisions are still exceedingly rare. This is even more true of asteroids, even though there are millions in our solar system. Remember in Empire Strikes Back where C-3PO said "the chance of successfully navigating an asteroid field is approximately 3,720 to 1"? He'd have been closer if he'd said "the chance of hitting anything is 3,720 to 1".

[142] I.e., to do with the body, such as medical risks, but also things like long term exposure to the space environment.

What's an Astronaut Worth?

However, before we consider the individual risks, we should look at how we approach the idea of risk. Humans are adapted to living in small bands on the plains of Africa, not the modern world. As such we carry some strange ideas about risk.

First of all, humans are very bad at comprehending large numbers. Though we may mathematically understand them, actually visualizing them is very difficult. How big is a million? Can you actually picture just how big that number is, let alone a billion? There are now over 7 billion (7,000,000,000) people on Earth. That's a pretty big number. By contrast, when our brains were evolving, we lived in groups of not more than 1000, and usually fewer than 200 people[143].

The fact is that well over 50 million people die in the world each year, and we rarely give this a second thought. This is equivalent to 137,000 deaths per day, or 1.6 deaths per second. Our brains really are geared based on that oft-quoted Stalin observation that "one death is a tragedy, a million deaths is a statistic". It is certainly possible to personalize one death - we can all imagine someone we know dying. On the other hand, the idea of a million deaths is simply unfathomable. We tend to make judgements and decisions based on these biases[144]. That's not to say that

---

[143] Interestingly, research has suggested that this range seems to be about the maximum number of "friends" people can maintain, including celebrities. This means that some of us actually think that Justin Bieber is part of our tribe.

[144] I suspect that loss aversion bias and emotional bias also play a big role in how we see things like a one-way Mars mission. The loss aversion bias means that people strongly prefer avoiding loss to making gains. This means that people are more likely to consider what they would give up on Earth than what they would gain by going to Mars. The emotional bias means that humans tend to weigh emotional considerations over purely rational ones.

these decisions are completely wrong – humans are after all, emotional creatures, and we can't make decisions without considering this. However, if we always put safety first, we would never accomplish anything. A reasonable balance must be struck.

On a personal level, we happily take more risks than we are conscious of every day, with no idea what the future holds. Yes, if we go on a risky space mission, we could die. But every day we live, we face the possibility of life shattering events. We could die in a car accident or be diagnosed with fatal cancer[145]. Each day we live is a gift, and we ought to make the most of it. Significantly, whenever there is a fatal accident in space flight - for example during the Apollo fire, or Challenger or Columbia accidents - though we deeply mourn the loss, we nonetheless consider the cause to have been worthy. Gus Grissom, who perished in the Apollo 1 fire in January 1967, had himself asserted that "the conquest of space is worth the risk of human life".

If we aren't taking risks, we aren't pushing hard enough. Gene Cernan, the last man on the Moon, observed: "I don't know if we have the mentality - I don't want to say guts - to take the kind of risks we did to get to the Moon the first time." Perhaps it is true that NASA today is incapable of taking the necessary risks. Perhaps it is because NASA exists in a political climate where we punish failure far more than we reward success. A NASA administrator's job appraisal seems to be based on maintaining a steady course, not losing anyone on a mission, and not making too many political waves.

The entire notion that it is responsible to play it safe when it comes to human spaceflight is flawed. This is not only because it fails to achieve

---

[145] Humans are notoriously bad at evaluating risk. Your chance of dying in a car accident is 1 in 100, at least 200 times higher than dying in an aviation accident. Yet, more humans are afraid of flying in airplanes. Your chance of dying of heart disease is 1 in 5. This makes donuts 20 times more dangerous than automobiles, but very few people are afraid of them.

the objectives for which it is intended, but also that it is a losing proposition in terms of human life. Sending humans to space is very expensive. If we aren't going to take risks that could result in accomplishments commensurate with the cost, we are not only throwing money away, we are throwing lives away.

Though it is a sad reality, the government must unfortunately make decisions all the time regarding how much money to allocate to one program or another in order to offset risk to human life. As Robert Zubrin has pointed out, for every $3 million spent on highway repairs, the US Federal Highway Administration estimates that it saves a life. For the $6 billion per year spent on human spaceflight, we could save 2000 lives. This price may be acceptable if human spaceflight actually opens up the solar system and helps build a better future for all, but it is illogical and immoral to spend vast sums of money just to save the lives of a few astronauts.

What about the ethics of a one-way mission? Is it unethical to send people to somewhere where we can't bring them back? My view is that as long as people understand what they are getting into and agree to it, how could it possibly be unethical? As with medical matters, informed consent and personal liberty should trump some conventional notion about what risks and sacrifices other people should be willing to tolerate. I sometimes even wonder if it's even ethical to bring people back from Mars. Are we even accomplishing our objective if people go out to Mars, and then turn around and come home?

Sending people one-way accomplishes all of the objectives, while a return trip dramatically escalates mission cost to the sum of hundreds of billions of dollars. Provided the crew volunteers, is it ethical to charge the public these sums just to bring a few humans home[146]? To think that

---

[146] I think an important distinction must be highlighted here between the Mars One plan and a suicide mission. The fact that humans who went to Mars would

it would be worth it strikes me as an elitist and illogical product of human thought biases. This enormous investment has a real cost, and the trade-off of bringing people home is certainly not favourable in either dollars or lives.

---

live a full and useful life is an essential part of my thinking. While I'm not certain that a suicide mission with volunteers would actually be unethical, I would not personally volunteer for such a mission – unless perhaps I had a terminal illness or was approaching the end of my life.

What's the Hurry?

So what's the rush? Why shouldn't we keep flying space station missions to reduce the risk, and wait until we have better technology to go to Mars?

Mars isn't getting closer. We had a better rocket in 1970, and while the space station program has yielded some life support technologies, these could have been developed on a much shorter time scale with greater incentives. The main technological advances in the past 40 years have been in computing, but this has a minor impact on human space mission, the computers constituting only a few percent of the overall mass[147].

The bottom line is that unfocused technological development isn't really making us better prepared for a mission to Mars. The most relevant technologies here are propulsion, but the nuclear rockets under development in the 1960s offered far more promise than anything on the drawing board today. While there have been some advances in electric propulsion, this is mostly useful for very small robotic spacecraft - it simply doesn't have enough thrust to drive a human mission.

Reducing the risk of space exploration too much can actually be counterproductive. Bertrand Russell's 1959 statement is noteworthy: "the future possibilities of space travel could show to the young that a world

---

[147] The first digital computer was developed to land humans on the Moon, but spacecraft tend to have laughably old computers due to concerns about reliability. The Space Shuttle's original computers were about 5000 times slower than today's smartphones, and were loaded using magnetic cassette tapes. Even after the upgrade in 1990, the shuttle computers has about 1/1500th the speed of a smartphone. In fact, this aversion to innovation in favor of older but proven technologies for is common in the space industry.

without war need not be a world without adventurous – and hazardous – glory".

For many[148], part of the attraction of an adventurous undertaking is the danger; not unnecessary reckless danger, but danger integral to doing something bold that has never been attempted before. Perhaps as a sad commentary on media coverage and the human psyche, the space program seems to receive a jolt of vitality for each close call. Witness the news coverage for Apollo 13 – that mission did far more for NASA's image than if they'd actually landed on the Moon as planned. By contrast, having people in space all the time, and sending people on regular flights to Earth orbit, has precisely the opposite effect. Too many low achievement missions erode popular support and create the dangerous illusion that going to space is routine.

No matter how thorough the planning, there will always be risks. Just going to space means that we need to launch humans on a rocket loaded to the brim with explosives, and bring them from 25,000 km/hr (16,000 mph) to zero in 30 minutes, generating temperatures of 1600 °C (2900 °F). As Columbia, Challenger, and several close calls with Soyuz underscore, this is dangerous business.

Since we are taking risks sending people to space, isn't it essential that we make sure to maximize their usefulness? Flying people to the space station once every five months for the next ten years will involve 24 flights, with 4 people each. If each flight carries a risk of 2%[149], this means we can expect to lose, on average, 2 people during that time. By comparison, a single Mars mission would have to have a failure rate of 50% to be so dangerous. You can account for an awful lot of launch failures, mechanical failures, and landing failures before you would come anywhere near a 50% risk rate. In the meantime, we'll be happily

---

[148] Myself definitely NOT included here.
[149] The approximate failure rate of the Space Shuttle.

accepting far more bone and muscle loss, and many times the cumulative radiation dose[150]. Can anyone doubt that we'd learn a heck of a lot more by going to Mars?

---

[150] Some might argue that it is more ethical to spread out the risk amongst many people even if the net total is higher. I'm not sure why this should be the case. Provided it is voluntary, is it more ethical to subject 100 people to a 1% risk than ten people to a 10%? I don't think so, although I suspect that the biases wired into our brains might say "yes". I also realize that it's possible to get caught up in the numbers and that this kind of moral equivalency can be dangerous. But I don't think this is the case here for voluntary assumption of risk.

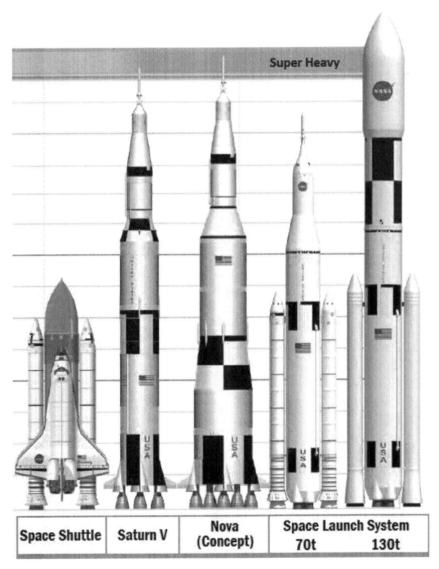

Comparison of NASA rockets from the past and today, including the Saturn V, Space Shuttle, and the Space Launch System (SLS) currently being designed

Mechanical Failure

"Houston, we have a problem". These dreaded and now famous words were spoken by Apollo 13 commander Jim Lovell on April 13, 1970[151]. En route to the Moon, 320,000 km (200,000 miles) from Earth, the crew heard a "loud bang", accompanied by fluctuations in electrical power. Initially, the crew thought they might have been struck by a small meteoroid. In fact, one of the two oxygen tanks in the Service Module had exploded after damaged Teflon insulation allowed two wires to short circuit, generating a spark. The shock also managed to knock out two fuel cells, and partially ruptured a line from the remaining oxygen tank, causing their remaining oxygen supplies to leak out into space over the next couple hours.

This was an utter disaster, knocking out all life support systems across the entire ship. If it hadn't been for the lunar module[152], the crew would surely have perished. As it was, using the lunar module as a lifeboat, the crew were barely able to struggle back to Earth.

Each Apollo spacecraft had over five million parts. An Apollo 13-style multi-system failure is probably the most dangerous part of a mission to Mars. Out in deep space, over a year from a potential return to Earth, such an accident would clearly spell disaster. The best solution to this problem is to always carry two copies of everything critical[153]. However, this can only go so far. Apollo 13 actually did have two separate oxygen

---

[151] Of course it was April 14 in other parts of the world, and no, it was not a Friday. It was a Monday. And if you listen to the actual voice recording, he said "we've had a problem", and Jack "Rusty" Swigert said it first.
[152] AKA "Lunar Lander"; i.e., the ship they would have used to get down to the surface of the Moon.
[153] In the space industry, this is called "redundancy".

tanks, yet the single explosion managed to knock out both, along with two power generators.

In deep space, this is always going to be a problem[154]. However, on the Martian surface, this will be far less of an issue. While we can't rule out the possibility of a total systems failure – for example a giant explosion – using multiple launches, we could essentially provide as many copies of whatever we'd like. If humans are that far away from Earth, we'd better make sure we have at least two, and hopefully three or more copies of everything.

We're still left with the possibility that the crew might not link up with their supplies, and that some supply flights would fail, but that's why you send multiple supply missions. Since you can only launch to Mars every two years or so, you'd have to make sure that the crew would have emergency backup supplies until the next possible resupply mission. However, none of these dangers are truly unique to a one-way Mars mission. In fact, without having to face the dangers of a deep space return flight to Earth[155], a one-way mission is probably safer.

---

[154] I suppose one solution would be to carry an entire spare spacecraft along for the ride. Say, that's not a bad idea!

[155] Not to mention takeoff from Mars – something we've never done before – with equipment that has been sitting in space for a number of years.

Humans in Space

Fears of what would happen to humans in unfamiliar environments have always been a concern. When we first started launching creatures to space in captured German missiles in the late 1940s, doctors came up with a long list of ways they thought humans would die when cut loose from the Earth[156].

Did organs rely on gravity to function? Would eyeballs change shape and render the astronauts blind? Would the heart just churn their blood? Would people be able to swallow[157]? Would they develop pneumonia, muscle cramps, or total nervous system failure[158]? It seems like we're back to duck, sheep, and rooster taking a ride in a balloon. Though a

---

[156] Therefore we sent monkeys, although the first animals in space were actually fruit flies in 1947, more than a decade before the famous dog Laika orbited the Earth in Sputnik 2. Laika was actually the third Soviet dog in space. She must have had a good PR contact (but not a very good travel agent considering she died). Six monkeys named Albert (I-VI) were flown between 1948 and 1951 – only one survived. The prestigious honor roll of American monkeys in space also includes Gordo, Miss Baker, and Miss Sam.

[157] Apparently, birds do have trouble swallowing in space - so no chickens on Mars. Well, you could put them on a centrifuge or feed them intravenously, but they probably wouldn't like that.

[158] As with the animals in the balloons, these concerns seem pretty silly to me. After all, humans are perfectly capable of hanging upside down without any trouble, and this is an even bigger gravity difference. For the record, David Blaine holds the record for hanging upside down the longest: 72 hours. While we're on the subject of weird space stuff, Total Recall got it wrong: you head wouldn't explode if you were exposed to space without a suit. You would lose consciousness in less than a minute and suffocate though. It is possible to explode from a pressure change, but you need a really big one, say from deep-sea diving. It happened in 1983 during the Byford Dolphin incident (don't look it up!). The pressure at the bottom of the deepest part of the ocean is equivalent to ten Polar Bears dancing on a postage stamp.

healthy dose of caution regarding the unknown is wise, these fears seem overblown, and are still with us today[159].

It is true that humans are not adapted to a space, or even a Mars, environment[160]. And there are real and serious health effects from being in space for long periods of time. The most sinister long-term effect is probably radiation, which will be discussed in more detail later. There are body fluid shifts, and significant changes in sensory organs like the inner ear, which normally relies on gravity to tell us which way is up. And finally, our bodies rely on the everyday resistance we encounter on Earth. A long trip in space causes serious degradation of your bones, muscles, and cardiovascular system.

However, apart from radiation, these changes are actually natural adaptations to being in space. Your body conserves resources, and if you don't use it, you lose it. Apart from radiation, there is no real danger of being in space and losing muscle and bone – you don't need them there. There is of course a real danger of sustaining accidental injury when returning to Earth. For the first few days back on Earth, as with the first few days in space, you feel very disoriented[161]. Your muscles are weak, and depending on how long you've been up there, you might have trouble walking or even standing. This muscle weakness also makes it harder to take emergency action when subjected to large G forces during re-entry. If you've been away for a really long time, you risk breaking bones back on Earth.

---

[159] This might be part of another cognitive bias, where humans have a tendency to overemphasize negative or scary news, and underemphasize positive news. It's why terrorism works, and why newspapers sell.

[160] Although by sheer coincidence, Mars is the most efficient place for human locomotion. The Moon has too little gravity and the Earth has too much. I suggest we double the length of Marathons on Mars, just to be fair.

[161] Most people suffer from motion sickness, aka "space sickness", and vomit for the first few days in space, even when they aren't moving due to the confusing sensory signals – which is one reason they never schedule space walks early in a mission.

Most of these negative health effects eventually recover. Your balance should return to normal after a few days. Your muscles should recover after a few weeks[162]. Your skeleton seems to make at least a partial recovery after a few months to a year, although it may never be quite as strong as it was.

Still, in the context of a Mars mission, I fail to see why these are major problems. Mars has 38% of the gravity of Earth. This means that any humans landing on Mars, even after a long space journey, would effectively be supermen. We've had people in space more than double the time it took for people to get to Mars. A shift on the space station is about the same duration. We seem to lose about 1-2% of our bone mass per month in space, and although this translates into a somewhat higher loss in strength, they should still be strong enough to deal with a Martian environment. Strenuous exercise in space can go a long way towards preventing muscle loss[163], and there shouldn't be much trouble on that front either. Even if the crew needs to rest and recover for a few days on the Martian surface to recover from disorientation, why should this be a problem?

Bone and muscle loss will probably continue on Mars, though at a much slower pace. With at least some gravity, exercise regimes on Mars should be much more effective. Moreover, there is a lower limit to how far bone and muscle loss will go. Humans, even floating in space, will not become bags of mush. Paraplegics typically lose 30% to 50% of their bone and muscle mass with no use whatsoever. True, we don't yet have any direct evidence of what partial gravity like that on Mars will do to the human body. However, it seems pretty evident that bone and muscles would

---

[162] I'm amazed as just how versatile our bodies are, considering they adapted in an Earth environment. After all, the creatures we evolved from over millions of years had no particular plans to go out and explore space. Although the bacteria they evolved from might have come from space.
[163] But not bone loss for some reason. Scientists have speculated that this might be due to a particular type of vibration, but the jury is still out.

plateau below a level adapted to Earth, and above a level adapted to space (or complete lack of use) – in fact, a level consistent with living on Mars.

This is perhaps another reason why a one-way mission to Mars makes a whole lot of sense. Going from a high gravity environment like Earth to a low one like Mars shouldn't present a problem. However, if we lived on Mars and were talking about exploring the Earth, we'd have serious problems. Humans returning to Earth after spending a long time in space, or even on the surface of Mars, might be so weak that they couldn't stand, would take quite some time to recover, and might never fully recover[164]. They might always be at risk of breaking bones: extremely osteoporotic women sometimes break bones just by shifting weight.

In terms of a return to Earth, is it ethical to accept these potentially debilitating long term health effects? As long as it is voluntary, this seems to me to be a personal choice, a matter of informed consent. People who understand the risks must make the tradeoffs for themselves. Moreover, I fail to see why it is more ethical to subject people to cumulative doses of these effects far in excess of a Mars mission by flying them in Earth orbit.

Over the next 10 years, we plan to fly at least 40 person-years of spaceflight (10 years times 4 crew on ISS). Over this time, although we are going to spread the damage over more people, the cumulative damage would be over 3 times higher than for a 3-year mission to Mars. Why is it more ethical to damage more people, if each person is damaged

---

[164] What would happen to a baby born on Mars? (Assuming this is possible, although I strongly suspect it would be). As a creature evolved on Earth, but having developed on Mars, I suspect they would be stronger than they needed to be for Mars, but weaker than Earthlings. They might never be able to fully adapt to an Earth environment – and would never make the cut for the Terran Olympics.

a bit less[165]? It's not even just the space station we're talking about. NASA conducts a large number of "bed rest" studies, where they pay people to do nothing except lie in bed and deteriorate, and observe the impact. Since bone and muscle disuse on Earth is a good simulation of being in space, this gives us some useful information about long duration spaceflight. Assuming it's all voluntary, I'm not sure there's an ethical problem with any of this. However, surely it must be more ethical to subject people to negative health effects only when they are unavoidable – for example on a mission to Mars[166]?

Something else that comes up a lot is medical care. On the surface, it seems perfectly normal to demand access to the best medical treatment we have available in the 21st century. However, humans have lived for over a million years, and we've had modern medical care maybe a hundred, depending on how you define the term. So for 99.99% of our existence we got along without medical care. Yes, we weren't especially healthy back then, but on a mission to Mars we'd have access to modern medical knowledge, if not always ideal treatment.

The mission would clearly carry basic emergency supplies. Additionally, all crew would be trained in first-aid, probably along with some basic emergency medicine training. A full medical doctor or surgeon might be part of the crew, but even if not, the crew would have access to medical

---

[165] One answer to this could be that the impacts are highly non-linear. For example, I think we can agree that it would be a lot more ethical to take a half-liter blood sample from 12 people than drain someone of their entire 6-liter supply. But I'm not sure that's the case here, at least not for the most part. This is certainly not the case for radiation, to be discussed next.

[166] I should probably also mention artificial gravity here, which can be generated by spinning the entire spacecraft, a small section of it, or something like a bed inside the spacecraft – like you see in 2001: A Space Odyssey. While I'm all for artificial gravity and would like to see it on a Mars mission, I don't think not being able to use it should be a reason not to go.

advice from Earth[167]. Yes, people might still die from conditions that can be cured using 21st century medicine on a Mars mission. But this happens all over the world today in remote, impoverished, and disaster-stricken regions – why should we demand a higher standard on Mars than we do on Earth? More to the point, if we are doing something that has never been done before, shouldn't we be willing to accept a higher level of risk?

---

[167] Telemedicine (where doctors perform operations remotely) would probably be of limited use due to the time delay.

Radiation

Radiation is scary. It can't be seen, smelled, felt, or otherwise detected until it's too late[168]. There are both short-term effects of high exposure[169], and long-term effects of lower doses. The problem with radiation is that it blasts through our bodies near the speed of light, causing all kinds of mayhem - primarily by knocking out pieces of DNA. Our bodies have a remarkable capacity for repairing DNA, which sustains damage from all kinds of causes, including environmental agents, toxins, and mistakes in ordinary cell division. Unrepaired mutations that pile up can lead in the long term to certain types of cancer. Certain people are at higher risk of developing these kinds of radiation-induced cancers: young women are most at risk, while older men are in less danger. Since cells divide much more slowly as you get older, many elderly can live years with various stages of cancer development without even noticing.

Although radiation is all around us all the time, luckily the Earth protects us from most of it[170]. There are several types of radiation to be found in space. Cosmic rays are high energy particles fired by supermassive black holes and supernovae. These, along with gamma rays and X-rays are mostly filtered out by collisions with nitrogen and oxygen molecules the Earth's atmosphere. There are also solar flares, which are released in

---

[168] Except when using technological devices like Geiger counters, of course!

[169] You don't want these for sure. Symptoms of a high radiation dose include headache, fever, nausea and vomiting, diarrhea, bleeding intestines, and total nervous system incapacitation. Call your doctor immediately if you experience any of these symptoms.

[170] Or maybe not so luckily: if we had evolved in a higher-radiation environment, we would be able to tolerate more radiation and wouldn't be talking about it now.

massive explosions from the Sun[171]. These would bathe the Earth in huge amounts of energy, if it weren't deflected by the Earth's magnetic field[172].

Radiation poses some challenge to Martian settlement. Solar flares release larger but slower moving particles, and are easier to shield against. This is a very good thing, because they would otherwise be by far the most dangerous type of space radiation. The spacecraft structure offers some protection, but water and organic matter provide even better shielding against this type of radiation. Depending on normal Sun activity cycles, small events occur every few days, with large events only rarely. Since they can be predicted several hours in advance and last a couple days at most, the best solution to deal with solar flares is to provide a "storm shelter" surrounded by layers of reserve water, wastewater, food stores, and waste of human origin[173].

Unfortunately, cosmic rays are so small and move so fast that they require too much shielding to be practical on a spacecraft. In fact, metal shielding can sometimes make the situation worse, because you can get secondary and more dangerous particles being fired off as cosmic rays smash into metal nuclei. Fortunately, whereas the radiation from solar flares could kill you quickly, cosmic rays are far less harmful in the short term. Still, cosmic rays represent about 95% of the radiation dose a Mars mission would receive.

---

[171] Everything about the Sun is massive, but these really deserve the term, being the equivalent of 160,000,000,000 megatons of TNT. Incidentally, you could fit 1.3 million Earths in the Sun, but there are much more massive stars. In the constellation Orion, there is a star called Betelgeuse that could swallow a billion Suns – or over 1.3 quadrillion Earths (1,300,000,000,000,000)! And it's not even close to the biggest.

[172] There is also ultraviolet radiation, absorbed by the ozone layer, but it doesn't seem critical here.

[173] If you know what I mean.

On the surface of Mars, exposed crews would receive about the same dose of cosmic rays as they do on the International Space Station, because this orbits above Earth's atmosphere. This level is about half as much as in deep space, because for ISS, Earth itself blocks half the cosmic rays, and the same would be true for Mars. Actually, it would be a bit less than that because the Martian atmosphere does filter out some of the cosmic rays, though not nearly as well as Earth.

So how much radiation would a Mars mission receive? According to data from the Radiation Assessment Detector (RAD) aboard NASA's Curiosity rover, a crew going to Mars and back would receive a round-trip radiation dose of about 0.66 Sievert[174]. What the heck does that mean? According to the Committee of the Biological Effects of Ionizing Radiation[175], an exposure of 1 Sievert equates to something roughly 50-165 deaths per million persons per rem during the first 25-27 years after irradiation.

Using the high end of these numbers, this suggests that a Mars mission would give you an additional 1.1% chance of getting cancer sometime later in your lifetime. NASA suggests these numbers could be higher, more like 3%, based on a backwards extrapolation. Risk of cancer varies by person and is not linear with exposure, so it's hard to tell exactly what the risk would be, but it's pretty safe to say that the risk to an individual would be somewhere between a 1% and 3% chance of developing cancer later in life.

NASA sets an astronaut's career limit radiation dose on a sliding scale that varies by age and gender. Older males are permitted the largest doses (1.47 Sieverts), and younger females are permitted the smallest

---

[174] 0.66 Sievert = 66 rem in case you're not a fan of Maximillian Sievert, or just don't like Sweden. For comparison, some workers at the Fukushima nuclear plant took this much radiation per hour.
[175] "The Effects on Populations of Exposure to Low Levels of Ionizing Radiation", 1972.

(0.47 Sieverts). Most other space agencies[176] set a fixed limit of 1 Sievert. Based on these limits, for some individuals like younger females, a mission to Mars would exceed their regulated lifetime dose limits.

But these are modern limits, established arbitrarily. In 1970, the National Academy of Sciences Space Studies Board made recommendations for long term space missions. They set a "primary reference risk" at the natural probability of developing cancer, or around 20%. Depending on age and gender, this would equate to an allowable radiation risk dosage of around 3-5 Sieverts[177] – 4 times higher than current limits. The Board noted that this was not intended to be a limit, but a reference. Higher limits could be tolerable for deep space missions, while lower limits should be imposed for things like space station operations.

An additional 20% chance of developing cancer in life may sound high, but we must keep in mind that this is roughly the risk assumed voluntarily by millions of people around the world today, who do so for the dubious pleasure of smoking. It also seems a bit strange that NASA is so concerned with radiation exposure on a Mars mission, when the current spaceflight plan has astronauts flying to the space station over the next 10 years, receiving a larger overall cumulative radiation dose[178].

Why are the NASA limits so low? Is this a case of government overregulation gone wild? It does strike me as rather odd that we should decide not to allow interplanetary explorers to voluntarily accept the same risk as smoking! Whatever the reason, it seems obvious, at least to

[176] Russian, European, Japanese, and Canadian.
[177] This is around the level of radiation taken by civilians in the vicinity of the 1986 Chernobyl plant. They do have an elevated risk of cancer, but this was not usually apparent for many years.
[178] This gets back to the question of if it's better to subject 96 astronauts to a 1% increased cancer rate over the next ten years than 4 astronauts to a 3% increased cancer rate. I don't see why this should be so.

me, that even if we want to impose safety standards for "routine" operations such as shifts on the space station, there should be much more flexibility for the first missions to other worlds. After all, in the entire history of human spaceflight, there is precisely zero evidence of elevated cancer rates amongst former astronauts. Moreover, since the base risk of developing cancer is around 20%, a 1% or even 3% increase really doesn't seem like very much.

Since we're talking here about the risk of developing cancer 25 years or more down the road, there is pretty much zero chance of radiation exposure having any mission impact. Thus, radiation in a Mars mission doesn't really represent a significant danger, as compared with the dangers of launch, landing, and mechanical failure. If you're worried about the risks of a Mars mission, radiation should not be ignored, but it shouldn't be at the top of your list of concerns either.

What about for a one-way Mars mission? One of the really nice things about a one-way mission is that since radiation doses in space are over double those on the surface even without any additional shielding, radiation is less of a short-term problem as compared with a two-way mission. The 210-day journey amounts to a radiation exposure of 0.38 Sieverts. This exposure is below the upper limits set by all space agencies for all ages and genders. On the Martian surface, shielding can be provided by several layers of soil – five meters of soil would provide approximately the same protection as the Earth's atmosphere. However, since natural background radiation on Earth is quite low (around 0.003 Sieverts per year) it shouldn't be necessary to use a covering this thick.

When Martian settlers are out exploring the surface, they will receive a higher dosage than on Earth, but somewhat less than on ISS. A settler who spent three hours per day, five days per week outside the habitat would receive a dosage of less than 0.02 Sieverts per year – about the same dosage as the limit imposed for radiation workers. If a settler moved from Earth to Mars at the age of 35 and lived to 75, this would represent less than a 4.5% increase in the chance of getting cancer in their lifetime (1% from the spaceflight, 3.5% from explorations on the surface)[179]. Surely, this level of risk, a level we already accept today in certain industries, should be considered acceptable for the colonization of a new planet.

---

[179] What about children born on Mars? Well, assuming this is possible, pregnant women should probably not go outside much. The FAA recommends no more than a 0.001 Sievert limit for pregnant airline pilots (who also absorb more cosmic rays than usual). This is equivalent to 18 days of Martian exploration. Young children should also probably limit time outside habitats. In any case, a Mars base in the early years probably wouldn't be a great environment for children.

The Mars Environment

The surface of Mars is a hostile place for humans. The pressure, at less than 1% of Earth's, might as well be space. Though the night and day on Mars are very similar to Earth's, the Sun is about half the size in the sky, and Mars is very cold. Even though in the very long run it might be possible to transform Mars back into a much more hospitable place, living on Mars is going to be a challenge for humans for the foreseeable future. Most of your time would be spent indoors, in a controlled pressurized environment, and it would be essential to make sure there was as much comfortable living space as possible.

When outside on excursions, you would have to wear a space suit and bring your oxygen supply at all times[180]. Current space suit designs are essentially bulky balloons filed with pressurized gas. This makes them clumsy and tiring to move around in, although this is somewhat offset by the low Martian gravity, where humans would have an especially easy time covering large distances efficiently on foot. There are better, more mobile, tight fitting space suits in the works, but current models are difficult and time consuming to put on and take off[181]. Therefore, for the near future, we are probably reliant on the old style pressure suits, not so different from the ones Neil Armstrong and Buzz Aldrin wore on the Moon.

The fact that Mars is very cold actually makes less difference than you would think. Though Mars is as cold as Antarctica, much less heat would

---

[180] Except if and when you used a methane-fueled pressurized ground rover for more distant explorations. This would be something like a camper trailer – a small, mobile, pressurized habitation and laboratory.
[181] These are called "mechanical counterpressure" space suits, pushing directly on the body as a substitute for gas pressure.

be lost to the thin air. Still, the potential for heat loss to the cold Martian ground dictates that insulation would an important consideration in habitation design[182]. The oft-mentioned Martian dust storms are also less of a threat than they would seem. Since the air is so thin, even wind speeds of 160 km/hr (100 mph) - a hurricane on Earth - would feel like a mere wisp of wind. Dust storms on Mars haven't posed a major threat to our rovers[183], and they wouldn't to humans either. The main danger of a dust storm would be its ability to block out sunlight and cover solar panels, reducing electricity available at a Mars base. This means that backup generators are going to be essential, most likely powered by methane which can be synthesized on Mars[184].

Regardless, dust storms aren't going to pose a major threat to a well-designed Mars base that has a backup plan for auxiliary power generation and storage. Power restrictions will mean that non-critical systems might have to be shut down for a few days or weeks, and energy-saving measures might have to be taken such as dimming lights, rationing water and air, and limiting exploration, but as soon as the storm passes and the solar panels are dusted off (by humans or rovers), everything should return to normal. Perhaps the more dangerous threat

---

[182] The cold, thin air on Mars always makes me think about the cold, thick air on Saturn's moon, Titan. The temperature on Titan is a mere 93.7 K (-180° C, -292° F) and the air is one and a half times thicker than on Earth. Heat loss would be extreme to say the least! However, if you did live on Titan, because of the low gravity and thick air, you could strap on wings and fly like a bird. Seems like a fair compensation.

[183] Indeed our rovers have generally lasted far longer than designed to. The NASA Spirit and Opportunity rovers were designed for a minimum 90 day life on Mars. Spirit lasted about 1900 days, and as of writing Opportunity is still going after over 3000 days, closing on the 10-year mark.

[184] The best solution may be a small nuclear reactor. The Soviets/Russians designed a suitable 5 KW, 320 kg model called TOPAZ (a Russian acronym this time) that they have flown a couple times. Able to operate for 3-5 years, a single one of these would be perfect for this kind of task. Six models were transferred to the US for testing in the 1990s, but a deal to purchase two for $13 million went south due to anti-nuclear protests. Darn those activists!

here is the dust itself. Fine dust has a nasty habit of sticking to everything, and could damage equipment and clog machinery[185]. Properly designed equipment seals, air locks, de-static, and decontamination procedures would be essential.

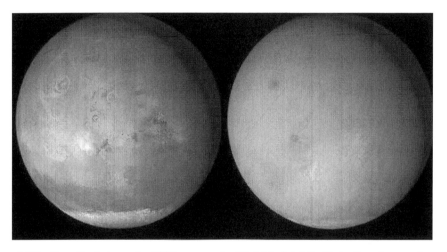

A Martian dust storm in 2001 as seen by Hubble (right). Note the obscured planetary features.

More than the dust itself, there might be a threat from chemicals in the dust called perchlorates. The global distribution of perchlorate salts on Mars is unknown; however, most of our robotic missions have found high concentrations[186]. We also find such high natural concentrations on Earth in environments that are similar to Mars, such as the Atacama desert in Chile. At around half a percent, perchlorates in the soil on Mars could be toxic to humans, interfering with iodide uptake in the thyroid gland.

The reason that perchlorates are dangerous is that they are highly reactive. Thankfully, this high reactivity also means that they are

---

[185] Not to mention give you a bad case of bathing suit itch.
[186] Specifically: Viking landers 0.1-0.9%, Pathfinder 0.55%, Spirit 0.06-0.68%, Opportunity 0.2-2.6%, Odyssey 0.2-0.8%, Phoenix 0.01-0.04%.

relatively easy to get rid of. It should be possible to decontaminate any exposed surfaces by reacting perchlorates with chemicals like sodium hydroxide, to generate water and harmless salts. The only real danger is if perchlorates contaminate food and water, and this should be preventable by adopting isolation and decontamination measures.

In the long run, the perchlorates might prove very useful. Ammonium perchlorate is a major component of solid rocket fuel – 70% in the case of the Space Shuttle solid rocket motors. When heated, calcium chlorate decomposes into calcium chloride solid, releasing oxygen. Additionally, this reaction produces a lot of energy, so perchlorates might provide an additional source of air and heat. Recent evidence has also revealed that archaea - ancient microbes unrelated to, but similar from our perspective to bacteria - can thrive on perchlorates. Thus, we may yet find evidence that cousins of these Earth microbes may have or may still thrive on the surface of the red planet.

Is it Really Unprecedented?

Two hundred years ago, if you were traveling from Europe to Australia[187], you would embark on a rickety sailing ship for a six month journey in extremely cramped, dank, and dark quarters. You might never be able to go above decks to get a breath of fresh air. Along the way, you'd face the dangers of pirates and war, storm and shipwreck, malnutrition and disease. The food would be horrible dried or mouldy "biscuit", the meat would usually be spoiled dried beef or salted fish, and you'd be lucky to get a few dried peas[188]. When you arrived, you'd have virtually nothing, and have to start building a life from scratch in an uninhabited environment[189]. The trip was dangerous and unforgiving, and life was tough.

It wasn't just Australia. If you're a North American with European ancestors, your relatives probably did something very similar, although the trip may have been a little shorter, with a typical trans-Atlantic voyage taking 4-6 weeks. In either case, they would have left friends and family, relatives and acquaintances, usually forever, with very limited

---

[187] Yes, you'd probably be a convict, but back then the most common offense for deportation was being in debt, so a lot of people in our society would be headed down under. And you could have been a soldier, because convicts need guards, and they usually went to Australia permanently too.

[188] But you might get a ration of rum to make the trip more tolerable. Unlike the US Navy which was dry from almost the beginning, the British Navy made sure their sailors were well stocked with a daily ration.

[189] Well of course there were the Aboriginies, but contact between Europeans and the Aboriginies of Australia didn't go so well. Aboriginies in the New South Wales region near Sydney were already fish farming, and seemed on the cusp of agriculture. But within a hundred years, they were virtually wiped out by the immigrating Europeans. Contact between natives and Europeans in Australia was perhaps even worse for the natives than in North America.

contact. Although you might write, the turn-around time for a letter was many months, and a high proportion of letters were lost. People back then understood these conditions and accepted them. They seemed normal.

These challenges faced by millions of overseas immigrants were normal, but humans have survived much worse in times of duress. Try to think for a moment what it might be like to be Anne Frank's family, hiding in a crawlspace, cramped with terrible conditions and little food, constantly afraid of being discovered by Nazi agents. Or imagine being stuck in a First World War trench, knee-deep in mud, surrounded by decomposing bodies, with rats scurrying about. Or how about being trapped at sea for months on end in a German U-boat during the Second World War? You'd be living in cramped quarters with terrible smells, never being able to surface, constantly being afraid of being attacked by allied patrols, never knowing if you'd make it home as the war was turning against you. Over 76% of all U-boat sailors who put to sea never returned[190].

I'm not suggesting that we be willing to accept these conditions – but I am suggesting that humans are more resilient than we give ourselves credit for, and we're able to tolerate extreme conditions if we consider them warranted. Admiral Richard Byrd, the famed Antarctic explorer confirmed this view: "Few men during their lifetime come anywhere near exhausting the resources dwelling within them. There are deep wells of strength that are never used."

I suspect that a one-way mission to Mars would seem perfectly normal to humans of the past. What has happened to us? Why do we balk at such a prospect? Why do we insist on controlled lives that we consider safe? Why do we insist that we can't do something bold and adventurous unless we have access to modern medical care? Has our high standard of

---

[190] Over 30,000 killed out of 39,000 who sailed from port. Any volunteers?

living in the modern, western world become an impediment, something that holds us back? A mission to Mars would probably be less dangerous than a sea voyage a couple hundred years ago. In the 1800s, the clipper ships were marvels of technology, the fastest sailing ships of the world, carrying valuable cargos around the globe. Yet 5% of clipper ships that left port were lost at sea, never to be heard from again.

A one-way mission to Mars only sounds unreasonable from a 21st century perspective. Most people look at the idea and think "I wouldn't have access to all that the 21st century has to offer". They generally look first at the downsides, and fail to consider what an amazing experience it could be, both on a personal level, and for our entire species.

A hundred years ago, the Antarctic explorer Ernest Shackleton put an advertisement in the paper seeking men for a "hazardous journey, [with] small wages, bitter cold, long months of complete darkness, constant danger, safe return doubtful, honor and recognition in the case of success." Shackleton had no trouble finding volunteers. Can you imagine stumbling across an advertisement like this in the New York Times today[191]? Yet, despite our modern civilization and all its comforts, we are still explorers deep down. We've done this before.

---

[191] I suppose you could if you've read about the Mars One Project!

Martian Psychology

It takes a bit longer than six months to get to Mars[192]. We've had people in space for much longer than that, so the actual trip to Mars isn't really unprecedented. Sure, there will be less living space than on the International Space Station, but conditions would probably be better than on Mir[193]. However, both for the long duration spaceflight, and for the long term stay on Mars, there will be some important psychological considerations. Indeed, the psychology of it all might be the most challenging part.

First of all, you're going to be living in close quarters with a very small number of people for a very long time. Even if new people arrive every two years, you're still essentially stuck with the same group of people forever. There have been some experiments on how humans might fare under these conditions. The Biosphere-2 experiments of the 1990s had seven or eight people live in a large simulated environment for two years. They had far more space than Mars One proposes and they were still on Earth, never in any real danger. Yet all crew members in the experiments agreed that the psychological issues were the biggest challenges.

---

[192] This varies between about 6 and 8 months because Earth and Mars go around the Sun at different rates so the alignment is not constant (and there is only an opportunity to launch to Mars every two years or so).

[193] Mir was notorious terrible conditions, smell, and fungus build-up. But you really do get used to the smell – it seems to plateau after a couple days. According to restricted hygiene studies conducted in the 1960s, where participants would spend weeks in the same clothes in cramped capsules, sometimes at elevated temperatures, it didn't get any worse after two weeks. That's reassuring.

We've also performed simulated Mars missions on Earth. The Mars 500 mission (actually 520 days) subjected a crew of 6 to a simulated trip to Mars in all relevant aspects between 2010 and 2011. The crew were locked in a small habitation module for the "outbound" and "return" trips, and were required to perform surface explorations in a "Mars" mock-up chamber when they "arrived". Effort was made to simulate every possible aspect of the trip (for example communication time lag), but of course it was not possible to simulate every aspect of a mission such as the risk or reduced gravity[194]. Though psychological issues were encountered, the mission was largely successful. The crew became lonely and bored, but this didn't result in huge problems. Every member of the crew reported that the experiment had not dissuaded them from an actual Mars mission.

The Mars Society has conducted simulations in the Utah desert and the Canadian arctic for over ten years at their simulated habitation. You couldn't ask for a more Martian environment on Earth than the Canadian far North: barren landscape, cold weather, no infrastructure, challenging logistics, and many similar geologic features, including impact craters.

These simulations have attempted to recreate every possible aspect of a Mars mission, including tracking supplies, always wearing space suits while outside the habitation, a communication time delay, and "Mars time", where personnel at the base operated on a 24 hour and 39 minute day[195]. A 365-day full year simulation is planned for 2014. NASA also

---

[194] I suppose they could have had the crew perform a bed rest study during the spaceflight portion of the trip, but they did not do this. It was a psychological experiment, not a physical one.

[195] This puts them off-schedule from "Earth" scientists they collaborate with, which is more problematic than your might think.

maintains a station in the Canadian arctic, and conducts 4, 8, and 12 month studies in Hawaii[196].

Most of these studies fail to represent what it would really be like to go to Mars. Replicating this is fundamentally impossible, because on Earth you are constantly aware that it's not the real thing. Usually there is little or no actual danger. For example, Mars 500 did an excellent job of reproducing the confinement of going to Mars, but not the true isolation from Earth. Still, they do provide some useful insights. For longer missions, including overwinters in Antarctica, we see a general a pattern where morale starts out high, drops considerably to a low point mid-mission, and then rises again with the light at the end of the tunnel.

It's hard to say what would happen when there is no light at the end of the tunnel. However, for most psychological issues like crew morale, there does seem to be a plateau. Provided you select the right individuals and intervene before serious problems arise, there doesn't seem to be a reason why humans couldn't handle going to Mars, even forever.

---

[196] Supposedly they chose Hawaii for the geology, but you've got to wonder if the weather has something to do with it.

The Right Stuff?

Who would be suited to go to Mars? Mars One gives its shopping list of characteristics as "resiliency, adaptability, curiosity, ability to trust, and resourcefulness". The ideal crew would be people who function well in a group. Having specific skills such as medicine, dentistry, engineering, or pilots skills may be useful, but not critical.

The ideal candidates would have to be able to get along well with others in any social dynamic. Confidence, resiliency, and an ability to maintain realistic expectations are critical to maintaining good morale. It's also much easier to select out than select in. A psychologist would look for any obvious weaknesses that would make a candidate unsuitable for long periods of isolation or to function in a group[197]. These characteristics could be either medical or behavioral[198].

People tend to underestimate what bothers them. You might be able to tolerate your annoying roommate blasting loud music on Earth, but what about in a confined spacecraft for six months when you can't leave and cool off? If you put enough pebbles on the scale, anyone can become agitated. People who would go to Mars have to be easy going in the extreme.

Isolation studies have shown that after a certain time – usually about six weeks – irrational antagonism starts to build. Even people who originally got along and who experienced no prior conflicts can develop irrational

---

[197] Personal correspondence with Sheryl Bishop – a social psychologist who has studied Mars mission simulations as well as expeditions to the Arctic, Antarctic, mountains, caves, and the Australian Outback.
[198] Basically, things like A) the Rorschach ink blot test: "is this person crazy?", or B) "Is this person just a jerk?" - or things like serious anxiety or depression.

feelings of annoyance and even aggression when confined together in an enclosed environment. Admiral Richard Byrd even preferred to carry out Antarctic overwinters alone in perilous conditions rather than risk this.

Astronauts who become frustrated always pick an outlet. Depending on personality, this is usually directed either towards each other, towards mission control, or towards themselves[199]. Stuck together for long periods of time, it is usually safer for crews to take out their tensions on mission control than each other. This can even build solidarity amongst the crew. Historically, mission controls have dealt with this by having an astronaut on Earth communicate with crews, because presumably someone who has been on a crew is in a better position to advocate on their behalf.

In a confined environment, psychological issues can become a big problem. How do you deal with someone who is depressed or aggressive? The goal is to address any emerging psychological problems before they become serious. Onboard remote monitoring by psychologists on Earth is critical, so that mission control and crewmates can intervene. Some tests focus on monitoring facial expression and voice pitch.

On a mission to Mars, robotic technologies like the Japanese "Kirobo" robot, launched to the International Space Station on August 3rd, 2013, may be useful for keeping astronauts company – and monitoring psychology at the same time. Direct questioning is a poor way to ascertain psychological state, but monitoring in daily routine and comparing facial expressions to a baseline for each subject has proven very effective in psychological experiments.

---

[199] This is especially the case with Japanese astronauts, who tend to become quiet and depressed rather than lashing out. But they do get to train using origami. Russian astronauts tend to blow off steam with outward competition bordering on aggression – it has occasionally come to blows. American crews most often lash out at mission control.

There is a trade-off between solitary and social people. Social people have an advantage because they are better suited to function in groups. However, solitary people who enjoy spending time alone can better cope with long stretches of boredom.

Someone who requires a lot of stimulation would go crazy - no "right stuff" test pilots here. More like easy going and friendly bookworms – people who would be satisfied with doing solo activities, but also empathetic in reaching out to their teammates. Not only did NASA look for a whole different type of candidate after the early adrenaline-filled missions, they even distinguished between Space Shuttle and Space Station crews. Anyone can last two weeks, but it takes the right kind of person to handle a six month rotation on the space station.

Younger generations may be both more and less accustomed to this kind of mission, let alone a mission to Mars. While people these days plugged into the latest gadgets may have shorter attention spans, they are often more comfortable sitting in front of a screen for days at a time playing video games. After all, much of the ordeal of going to Mars will be dealing with long periods of boredom.

Mixed-gender crews seem to perform best, leading to fewer fist fights, higher productivity, and better teamwork. Crews benefit from the competitiveness and focus that males tend to bring, and empathy and community that women tend to bring. All-male crews are capable of performing, but tend to lead to higher competitiveness, and more potential for aggression. All-female crews tend to perform the worst[200].

---

[200] In the words of space psychologist Sheryl Bishop, all women crews lead to a lot of "chit-chat".

As they had historically been from naval vessels[201], women were excluded from Antarctic bases until 1974, when a spinster and nun were first added to an overwinter crew. Since then, the numbers of women in Antarctica have been rising, and now comprise around one-third of crews. For long duration missions, some psychologists have suggested that the best crews would be comprised of non-monogamous couples[202].

---

[201] The US Navy is now integrated except on submarines. Still holding back from full integration, the Navy is now starting to "man" some submarines with all-female crews.

[202] Or even "emotionally compatible bisexual crews".

Earth Out-of-Sight

When Yuri Gagarin allowed himself to be launched on the tip of a giant rocket beyond the protective atmosphere of the Earth for the first time on April 12, 1961, no one was really sure what would happen[203]. Some psychologists thought that people might go crazy, unable to make rational decisions. In fact, to prevent Yuri from doing something that would jeopardize the flight, he was locked out of the flight controls on his Vostok capsule. As a hedge, he was given a combination to unlock the flight controls, sealed in an envelope, and only to be opened in case of extreme emergency – the theory being, if he had enough presence of mind to enter the combination, at least he wasn't insane. Thus the first human in space was merely a passenger, with no ability to influence the flight in any way[204]. The first flight in space was a major psychological achievement, demonstrating that fears of the unknown are often unfounded.

Going to Mars is different from going to space, but the main differences are time and distance. Earth will fade to a pale blue dot in the night sky, just as Mars appears to us a pale red dot, not so different from a star. More than on the International Space Station, crews will be beyond help, and beyond influencing events on Earth. Even on the space station, this has generated distress in certain individuals. For example, a death in the

---

[203] Again, since a dozen monkeys and 3 dogs had already been there, it seems awfully strange to think anything especially different would happen with humans. Yet, in the words of space psychologist Eugene Brody in 1959: "separation from Earth, with all its unconscious symbolic significance, might in theory produce something akin to the panic of schizophrenia".

[204] American propaganda used this fact to highlight the fact that while Soviet "automation" reduced the human to a cog in the machine, *their* pilots used actual flying skill in a demonstration of independent spirit – despite the fact that they were preceded by a parade of space primates.

family produces a sense of helplessness when trapped hundreds of kilometers from home, unable to comfort friends and family. Mir astronauts on even longer flights drew comfort from photographs of family, email contact, etc., but dwelling too much on the separation from Earth can make things worse.

Though the Apollo astronauts experienced a very small communications time delay from the Moon of about a second or two, it takes between 4 and 24 minutes for a signal to get between Earth and Mars. This will mean that far more than any space mission of the past, crews will be on their own for short-term decisions. At a certain point, any mission control on Earth will become a second opinion, and increasingly autonomous crews will make their own decisions, whether we like it or not. Ground controllers must accept some loss of power, relinquishing the parent-child relationship. The reality is that Earth will be able to influence decisions, but won't always get to call all the shots.

What about the very fact of leaving Earth forever? Would this necessarily lead to anxiety and depression? There are certainly some extreme sacrifices here. For example, crews would have to accept the possibility that they might never see nature again. Imagine never seeing a tree again in your life. On the other hand, there are many cultures on Earth who never see trees. Mongolian herders and Arabian Bedouin live in featureless regions. The Inuit have historically lived in barren tundra with no vegetation at all. Even today, most people never see jungles, except on television.

In these types of situations, I think the crew would learn to appreciate the little things. Imagine how lovely plants in a Martian greenhouse would seem, powerful reminders of Earth sights and scents. Interiors that featured a lot of variety, lots of colors, natural lines, greenery, and a welcoming atmosphere would go a long way towards offsetting depression.

So would humans go crazy on a long flight to Mars, or once on the surface? Some might, but if you select the right individuals, probably not. Sergei Krikalev, a veteran of six space missions who has spent more time in space than any other astronaut[205], scoffed at the idea: "Psychologists need to write papers. People used to think humans would go insane riding trains and seeing the trees whizz by".

Our first view of Earth from Mars, as captured by NASA's Curiosity Rover in early 2014.

---

[205] A total of over 800 days – more than 2 years.

Organizing Crews

In some ways, it seems a bit strange that Mars One has already begun to look for candidates to send to Mars. After all, the mission is far more contingent on funding and hardware considerations than finding a crew. However, this is, after all, a human story. Moreover, it really does take a lot of time to identify candidates, and prepare them for many years. The current Mars One plan is to select candidates in 2013 and 2014, and start initial training in 2015 for three months at a time.

Full time training would start soon after, in analogue environments like Antarctica. With something this big, it probably makes a lot of sense to start early and have time to test the waters, both for the psychologists who would assess candidates, and for the crews themselves. Perhaps it is not so strange after all – NASA started recruiting for the Apollo missions just after the decision was made to go to the Moon[206].

Mars One is looking for teams of four, most likely two males and two females. So how the heck would you assemble a team that would spend inordinate amounts of time together during training – and potentially the rest of their lives? We've faced something a little similar in the past.

When assembling bomber crews during the Second World War – people you fought with and might die with - each country had a different strategy. The Germans assigned crews based on what city or geographic region they came from – the idea being that people from the same area

---

[206] That brings up an interesting question – why not use current astronaut volunteers? I suspect that it is both because the skills required are not direct mappings and can be trained, and also that the idea that "anyone could go to Mars" is a big part of the intended appeal for a global audience. Mars One isn't really looking for astronauts after all, they are looking for settlers.

would have more in common. The Americans assigned crews from above by skill sets, and captains could put in special requests for particular individuals. Meanwhile, the British threw a big party, and whoever hit it off got to fly together.

However, assembling crews for Mars One would probably be based entirely on psychological assessment. People can be trained to build bases, grow food, repair rovers, and collect samples for scientific investigation, but it seems that not everyone can be trained to deal with harsh psychological challenges. Everyone has different psychological characteristics, and balancing these across a crew would probably be a key factor. You'd want the right mix of empathy, intuition, perception, resiliency, and emotional stability, in the right doses and combinations.

They might run a lot of psychological assessments, stick people together, test them in Earth-based training, and then mix it up. After enough-trial and-error, you'd get balanced teams for any situation. This would be essential: although funding and hardware development will be critical to ensure that the mission can be carried out, psychological crew balancing might well determine its success or failure in the long run.

Mars Calling

We have the technology to go to Mars within the next 10 years. Unlike many challenges we face, it's not so much new thinking that's required. We need a dose of old thinking, harking back to when we were willing to take majors risks for major gains. If we choose to go to Mars, doing it one-way shouldn't be anything to fear: it makes sense as the most logical and efficient way to proceed. Rather than waste decades on technology development to ensure the return of a few individuals to Earth, we should embrace our expansion into space as a mission of settlement from the start.

There are a lot of unanswered questions, but the only way to answer some is to brave the unknown. We have a lot of work to do, but we are much better prepared to send people to Mars now than we were in 1961 to send people to the Moon. Back then, we not only had to build all the hardware from scratch, but we had to develop many of the technologies too.

That being said, all of this is a very long shot. There are tremendous hurdles to overcome in both hardware and funding. Mars One is not an engineering company – it has no technical ability to get to Mars. It is a fundraising and organizational umbrella. In some ways, this is a good thing, because it can remain global, impartial, and free to purchase hardware from any vendor[207]. However, the plain fact is that Mars One is utterly dependent on hardware suppliers for every aspect of the mission. Within a few years, we should have a better idea of the technical

---

[207] The fact that multiple vendors may be involved might cause issues with hardware intercompatibility. Regardless, almost certainly the lion's share of the work would go to SpaceX. No SpaceX, no Mars – at least not yet.

feasibility of this kind of plan. Although we fundamentally understand how to send people to Mars, we don't have the rockets, interplanetary spacecraft, or Mars habitats sitting on the launch pad waiting for the signal. Unless NASA, SpaceX, or some other entity builds the spacecraft and performs the engineering work, no one is going to Mars.

Before anyone heads for Mars, we'll have to see multiple test flights, multiple Mars landings, and have a base positioned on the surface filled with backup emergency supplies. With proven hardware, and a base pre-established, none of this would look quite so risky. In fact, it would represent a level of preparation far in excess of when we first sent humans to the Moon.

However, it is also a tremendous amount to accomplish in a very short time. It is likely that the fate of Mars One is strongly tied to the planned (and more realizable) Inspiration Mars 2018 flyby mission. This 500+ day mission would take a male and female couple around the red planet on the longest spaceflight in history, beyond help from Earth. Unlike Mars One, Inspiration Mars is already backed by wealthy individuals, with a less ambitious objective. If Inspiration Mars were to succeed in accomplishing the first human interplanetary flight, it would prove that it is possible for humans to travel to other planets. Then, the next logical step of landing would seem like a relatively minor one.

The Mars One project is extraordinarily ambitious, and there are lots of dominoes that have to fall into place for it to work. However, there is nothing impossible about the plan. If a plan to send humans on a one-way mission to Mars were to be widely accepted around the world, it would most certainly happen. One path would be for a large organization like NASA to support it. Other paths to Mars could be overwhelming global grassroots support, or global success of precursor efforts on Earth, followed by strong investment, both big and small. A visionary goal with a near-term deadline is essential for progress to be made. Mars One offers this. Even if it were delayed, or failed completely, it could still provide a long-term stimulus that might take us to Mars.

Planet Earth: if you think it's time for humans to start expanding into space, it's up to you to make it happen.

This is your mission to Mars.

*　　*　　*

About the Author

Andrew Rader is an Aerospace Engineer with a Ph.D. in human spaceflight from MIT (2009) who has worked as a spacecraft systems engineer on half a dozen space missions.

In 2013, Andrew won Discovery Channel's top competitive television series, *Canada's Greatest Know-it-All*, and has appeared in many radio and television interviews.

He is currently a second-round candidate for the Mars One mission, but is otherwise unaffiliated with project, which aims to send humans to the Red Planet starting in 2023.

Andrew is an avid trivia player and public speaker, giving talks at schools, conventions, museums, and other venues. He is also a strategy game designer and producer of numerous educational science, science fiction, space, and history videos, which you can find on his Youtube Channel (*youtube.com/AndrewRader*).

www.andrew-rader.com

Twitter:

@marsrader (space/personal)

@weird_sci (science)

@weird_hist (history)

Made in the USA
Lexington, KY
28 March 2014